'It is as in the literary imagination climate change were somehow akin to extraterrestrials or interplanetary travel'

Uncanny and Improbable Events

AMITAV GHOSH

PENGUIN BOOKS — GREEN IDEAS

PENGUIN BOOKS

UK | USA | Canada | Ireland | Australia
India | New Zealand | South Africa

Penguin Books is part of the Penguin Random House group of companies
whose addresses can be found at global.penguinrandomhouse.com.

Penguin
Random House
UK

First published in *The Great Derangement* by The University of Chicago Press 2016
This extract published in Penguin Books 2021

002

Set in 11/13pt Dante MT Std
Typeset by Jouve (UK), Milton Keynes
Printed and bound in Great Britain by Clays Ltd, Elcograf S.p.A.

The authorized representative in the EEA is Penguin Random House Ireland,
Morrison Chambers, 32 Nassau Street, Dublin D02 YH68

A CIP catalogue record for this book is available from the British Library

ISBN: 978–0–141–99690–5

www.greenpenguin.co.uk

Contents

I.

Who can forget those moments when something that seems inanimate turns out to be vitally, even dangerously alive? As, for example, when an arabesque in the pattern of a carpet is revealed to be a dog's tail, which, if stepped upon, could lead to a nipped ankle? Or when we reach for an innocent looking vine and find it to be a worm or a snake? When a harmlessly drifting log turns out to be a crocodile?

It was a shock of this kind, I imagine, that the makers of *The Empire Strikes Back* had in mind when they conceived of the scene in which Han Solo lands the *Millennium Falcon* on what he takes to be an asteroid – only to discover that he has entered the gullet of a sleeping space monster.

To recall that memorable scene now, more than thirty-five years after the making of the film, is to recognize its impossibility. For if ever there were a Han Solo, in the near or distant future, his assumptions about interplanetary objects are certain to be very different from those that prevailed in California at the time when the film was made. The humans of the future will surely understand, knowing what they presumably will know about the history of their forebears on Earth, that only in one, very brief era, lasting less than three centuries, did a significant number of their kind believe that planets and asteroids are inert.

2.

My ancestors were ecological refugees long before the term was invented.

They were from what is now Bangladesh, and their village was on the shore of the Padma River, one of the mightiest water ways in the land. The story, as my father told it, was this: one day in the mid-1850s the great river suddenly changed course, drowning the village; only a few of the inhabitants had managed to escape to higher ground. It was this catastrophe that had unmoored our forebears; in its wake they began to move westward and did not stop until the year 1856, when they settled once again on the banks of a river, the Ganges, in Bihar.

I first heard this story on a nostalgic family trip, as we were journeying down the Padma River in a steamboat. I was a child then, and as I looked into those swirling waters I imagined a great storm, with coconut palms bending over backward until their fronds lashed the ground; I envisioned women and children racing through howling winds as the waters rose behind them. I thought of my ancestors sitting huddled on an outcrop, looking on as their dwellings were washed away.

To this day, when I think of the circumstances that have shaped my life, I remember the elemental force that untethered my ancestors from their homeland and launched them on the series of journeys that preceded, and made possible, my own travels. When I look into my past the river seems to meet my eyes, staring back, as if to ask, Do you recognize me, wherever you are?

Recognition is famously a passage from ignorance to knowledge. To recognize, then, is not the same as an initial introduction. Nor does recognition require an exchange of words: more often than not we recognize mutely. And to recognize is by no means to understand that which meets the eye; comprehension need play no part in a moment of recognition.

The most important element of the word *recognition* thus lies in its first syllable, which harks back to something prior, an already existing awareness that makes possible the passage from ignorance to knowledge: a moment of recognition occurs when a prior awareness flashes before us, effecting an instant change in our understanding of that which is beheld. Yet this flash cannot appear spontaneously; it cannot disclose itself except in the presence of its lost other. The knowledge that results from recognition, then, is not of the same kind as the discovery of something new: it arises rather from a renewed reckoning with a potentiality that lies within oneself.

This, I imagine, was what my forebears experienced on that day when the river rose up to claim their village: they awoke to the recognition of a presence that had molded their lives to the point where they had come to take it as much for granted as the air they breathed. But, of course, the air too can come to life with sudden and deadly violence – as it did in the Congo in 1988, when a great cloud of carbon dioxide burst forth from Lake

Nyos and rolled into the surrounding villages, killing 1,700 people and an untold number of animals. But more often it does so with a quiet insistence – as the inhabitants of New Delhi and Beijing know all too well – when inflamed lungs and sinuses prove once again that there is no difference between the without and the within; between using and being used. These too are moments of recognition, in which it dawns on us that the energy that surrounds us, flowing under our feet and through wires in our walls, animating our vehicles and illuminating our rooms, is an all-encompassing presence that may have its own purposes about which we know nothing.

It was in this way that I too became aware of the urgent proximity of nonhuman presences, through instances of recognition that were forced upon me by my surroundings. I happened then to be writing about the Sundarbans, the great mangrove forest of the Bengal Delta, where the flow of water and silt is such that geological processes that usually unfold in deep time appear to occur at a speed where they can be followed from week to week and month to month. Overnight a stretch of riverbank will disappear, sometimes taking houses and people with it; but elsewhere a shallow mud bank will arise and within weeks the shore will have broadened by several feet. For the most part, these processes are of course cyclical. But even back then, in the first years of the twenty-first century, portents of accumulative and irreversible change could also be seen, in receding shorelines and a steady intrusion of salt water on lands that had previously been cultivated.

This is a landscape so dynamic that its very changeability leads to innumerable moments of recognition. I captured some of these in my notes from that time, as, for example, in these lines, written in May 2002: 'I do believe it to be true that the land here is demonstrably alive; that it does not exist solely, or even incidentally, as a stage for the enactment of human history; that it is [itself] a protagonist.' Elsewhere, in another note, I wrote, 'Here even a child will begin a story about his grandmother with the words: "in those days the river wasn't here and the village was not where it is . . ."'

Yet, I would not be able to speak of these encounters as instances of recognition if some prior awareness of what I was witnessing had not already been implanted in me, perhaps by childhood experiences, like that of going to look for my family's ancestral village; or by memories like that of a cyclone, in Dhaka, when a small fishpond, behind our walls, suddenly turned into a lake and came rushing into our house; or by my grandmother's stories of growing up beside a mighty river; or simply by the insistence with which the landscape of Bengal forces itself on the artists, writers, and filmmakers of the region.

But when it came to translating these perceptions into the medium of my imaginative life – into fiction, that is – I found myself confronting challenges of a wholly different order from those that I had dealt with in my earlier work. Back then, those challenges seemed to be particular to the book I was then writing, *The Hungry Tide*; but now, many years later, at a moment when the accelerating impacts of global warming have begun to threaten the very existence of low-lying areas like the Sundarbans, it seems to me that those problems have far wider implications. I have come to recognize that the challenges that climate change poses for the contemporary writer, although specific in some respects, are also products of something broader and older; that they derive ultimately from the grid of literary forms and conventions that came to shape the narrative imagination in precisely that period when the accumulation of carbon in the atmosphere was rewriting the destiny of the earth.

3.

That climate change casts a much smaller shadow within the landscape of literary fiction than it does even in the public arena is not hard to establish. To see that this is so, we need only glance through the pages of a few highly regarded literary journals and book reviews, for example, the *London Review of Books*, the *New York Review of Books*, the *Los Angeles Review of Books*, the *Literary Journal*, and the *New York Times Review of Books*. When the subject of climate change occurs in these publications, it is almost always in relation to nonfiction; novels and short stories are very rarely to be glimpsed within this horizon. Indeed, it could even be said that fiction that deals with climate change is almost by definition not of the kind that is taken seriously by serious literary journals: the mere mention of the subject is often enough to relegate a novel or a short story to the genre of science fiction. It is as though in the literary imagination climate change were somehow akin to extraterrestrials or interplanetary travel.

There is something confounding about this peculiar feedback loop. It is very difficult, surely, to imagine a conception of seriousness that is blind to potentially life-changing threats. And if the urgency of a subject were indeed a criterion of its seriousness, then, considering what climate change actually portends for the future of the earth, it should surely follow that this would be the principal preoccupation of writers the world over – and this, I think, is very far from being the case.

But why? Are the currents of global warming too wild to be navigated in the accustomed barques of narration? But the truth, as is now widely acknowledged, is that we have entered a time when the wild has become the norm: if certain literary forms are unable to negotiate these torrents, then they will have failed – and their fail-ures will have to be counted as an aspect of the broader imaginative and cultural failure that lies at the heart of the climate crisis.

Clearly the problem does not arise out of a lack of information: there are surely very few writers today who are oblivious to the current disturbances in climate systems the world over. Yet, it is a striking fact that when novelists do choose to write about climate change it is almost always outside of fiction. A case in point is the work of Arundhati Roy: not only is she one of the finest prose stylists of our time, she is passionate and deeply informed about climate change. Yet all her writings on these subjects are in various forms of nonfiction.

Or consider the even more striking case of Paul

Kingsnorth, author of *The Wake*, a much-admired historical novel set in eleventh-century England. Kingsnorth dedicated several years of his life to climate change activism before founding the influential Dark Mountain Project, 'a network of writers, artists and thinkers who have stopped believing the stories our civilization tells itself.' Although Kingsnorth has written a powerful nonfiction account of global resistance movements, as of the time of writing he has yet to publish a novel in which climate change plays a major part.

I too have been preoccupied with climate change for a long time, but it is true of my own work as well, that this subject figures only obliquely in my fiction. In thinking about the mismatch between my personal concerns and the content of my published work, I have come to be convinced that the discrepancy is not the result of personal predilections: it arises out of the peculiar forms of resistance that climate change presents to what is now regarded as serious fiction.

4.

In his seminal essay 'The Climate of History,' Dipesh Chakrabarty observes that historians will have to revise many of their fundamental assumptions and procedures in this era of the Anthropocene, in which 'humans have become geological agents, changing the most basic physical processes of the earth.' I would go further and add that the Anthropocene presents a challenge not only to the arts and humanities, but also to our commonsense understandings and beyond that to contemporary culture in general.

There can be no doubt, of course, that this challenge arises in part from the complexities of the technical language that serves as our primary window on climate change. But neither can there be any doubt that the challenge derives also from the practices and assumptions that guide the arts and humanities. To identify how this happens is, I think, a task of the utmost urgency: it may well be the key to understanding why contemporary culture finds it so hard to deal with climate change. Indeed, this is perhaps the most important question ever to confront *culture* in the broadest sense – for let us make no mistake: the climate crisis is also a crisis of culture, and thus of the imagination.

Culture generates desires – for vehicles and appliances, for certain kinds of gardens and dwellings – that are among the principal drivers of the carbon economy. A speedy convertible excites us neither because of any love for metal and chrome, nor because of an abstract understanding of its engineering. It excites us because it evokes an image of a road arrowing through a pristine landscape; we think of freedom and the wind in our hair; we envision James Dean and Peter Fonda racing toward the horizon; we think also of Jack Kerouac and Vladimir Nabokov. When we see an advertisement that links a picture of a tropical island to the word *paradise*, the longings that are kindled in us have a chain of transmission that stretches back to Daniel Defoe and Jean-Jacques Rousseau: the flight that will transport us to the island is merely an ember in that fire. When we see a green lawn that has been watered with desalinated water, in Abu Dhabi or Southern California or some other environment where people had once been content to spend their water thriftily in nurturing a single vine or shrub, we are looking at an expression of a yearning that may have been midwifed by the novels of Jane Austen. The artifacts and commodities that are conjured up by these desires are, in a sense, at once expressions and concealments of the cultural matrix that brought them into being.

This culture is, of course, intimately linked with the wider histories of imperialism and capitalism that have shaped the world. But to know this is still to know very little about the specific ways in which the matrix interacts with different modes of cultural activity: poetry, art, architecture, theater, prose fiction, and so on. Throughout history these branches of culture have responded to war, ecological calamity, and crises of many sorts: why, then, should climate change prove so peculiarly resistant to their practices?

From this perspective, the questions that confront writers and artists today are not just those of the politics of the carbon economy; many of them have to do also with our own practices and the ways in which they make us complicit in the concealments of the broader culture. For instance: if contemporary trends in architecture, even in this period of accelerating carbon emissions, favor shiny, glass-and-metal-plated towers, do we not have to ask, What are the patterns of desire that are fed by these gestures? If I, as a novelist, choose to use brand names as elements in the depiction of character, do I not need to ask myself about the degree to which this makes me complicit in the manipulations of the marketplace?

In the same spirit, I think it also needs to be asked, What is it about climate change that the mention of it should lead to banishment from the preserves of serious fiction? And what does this tell us about culture writ large and its patterns of evasion?

In a substantially altered world, when sea-level rise has swallowed the Sundarbans and made cities like Kolkata, New York, and Bangkok uninhabitable, when readers and museum-goers turn to the art and literature of our time, will they not look, first and most urgently, for traces and portents of the altered world of their inheritance? And when they fail to find them, what should they – what can they – do other than to conclude that ours was a time when most forms of art and literature were drawn into the modes of concealment that prevented people from recognizing the realities of their plight? Quite possibly, then, this era, which so congratulates itself on its self-awareness, will come to be known as the time of the Great Derangement.

5.

On the afternoon of March 17, 1978, the weather took an odd turn in north Delhi. Mid-march is usually a nice time of year in that part of India: the chill of winter is gone and the blazing heat of summer is yet to come; the sky is clear and the monsoon is far away. But that day dark clouds appeared suddenly and there were squalls of rain. Then followed an even bigger surprise: a hailstorm.

I was then studying for an MA at Delhi University while also working as a part-time journalist. When the hailstorm broke, I was in a library. I had planned to stay late, but the unseasonal weather led to a change of mind and I decided to leave. I was on my way back to my room when, on an impulse, I changed direction and dropped in on a friend. But the weather continued to worsen as we were chatting, so after a few minutes I decided to head straight back by a route that I rarely had occasion to take.

I had just passed a busy intersection called Maurice Nagar when I heard a rumbling sound somewhere above. Glancing over my shoulder I saw a gray, tube-like extrusion forming on the underside of a dark cloud: it grew rapidly as I watched, and then all of a sudden it turned and came whiplashing down to earth, heading in my direction.

Across the street lay a large administrative building. I sprinted over and headed toward what seemed to be an entrance. But the glass-fronted doors were shut, and a small crowd stood huddled outside, in the shelter of an overhang. There was no room for me there so I ran around to the front of the building. Spotting a small balcony, I jumped over the parapet and crouched on the floor.

The noise quickly rose to a frenzied pitch, and the wind began to tug fiercely at my clothes. Stealing a glance over the parapet, I saw, to my astonishment, that my surroundings had been darkened by a churning cloud of dust. In the dim glow that was shining down from above, I saw an extraordinary panoply of objects flying past – bicycles, scooters, lampposts, sheets of corrugated iron, even entire tea stalls. In that instant, gravity itself seemed to have been transformed into a wheel spinning upon the fingertip of some unknown power.

I buried my head in my arms and lay still. Moments later the noise died down and was replaced by an eerie silence. When at last I climbed out of the balcony, I was confronted by a scene of devastation such as I had never before beheld. Buses lay overturned, scooters sat perched on treetops, walls had been ripped out of buildings, exposing interiors in which ceiling fans had been twisted into tulip-like spirals. The place where I had first thought to take shelter, the glass-fronted doorway, had been reduced to a jumble of jagged debris. The panes had shattered, and many people had been wounded by the shards. I realized that I too would have been among the injured had I remained there. I walked away in a daze.

Long afterward, I am not sure exactly when or where, I hunted down the *Times of India's* New Delhi edition of March 18. I still have the photocopies I made of it.

'30 Dead,' says the banner headline, '700 Hurt As Cyclone Hits North Delhi.'

Here are some excerpts from the accompanying report: 'Delhi, March 17: At least 30 people were killed and 700 injured, many of them seriously, this evening when a freak funnel-shaped whirlwind, accompanied by rain, left in its wake death and devastation in Maurice Nagar, a part of Kingsway Camp, Roshanara Road and

Kamla Nagar in the Capital. The injured were admitted to different hospitals in the Capital.

'The whirlwind followed almost a straight line . . . Some eyewitnesses said the wind hit the Yamuna river and raised waves as high as 20 or 30 feet . . . The Maurice Nagar road . . . presented a stark sight. It was littered with fallen poles . . . trees, branches, wires, bricks from the boundary walls of various institutions, tin roofs of staff quarters and dhabas and scores of scooters, buses and some cars. Not a tree was left standing on either side of the road.'

The report quotes a witness: 'I saw my own scooter, which I had abandoned on the road, during those terrifying moments, being carried away in the wind like a kite. We saw all this happening around but were dumbfounded. We saw people dying . . . but were unable to help them. The two tea-stalls at the Maurice Nagar corner were blown out of existence. At least 12 to 15 persons must have been buried under the debris at this spot. When the hellish fury had abated in just four minutes, we saw death and devastation around.'

The vocabulary of the report is evidence of how unprecedented this disaster was. So unfamiliar was this phenomenon that the papers literally did not know what to call it: at a loss for words they resorted to 'cyclone' and 'funnel-shaped whirlwind.'

Not till the next day was the right word found. The headlines of March 19 read, 'A Very, Very Rare Phenomenon, Says Met Office': 'It was a tornado that hit northern parts of the Capital yesterday – the first of its kind . . . According to the Indian Meteorological Department, the tornado was about 50 metres wide and covered a distance of about five k.m. in the space of two or three minutes.'

This was, in effect, the first tornado to hit Delhi – and indeed the entire region – in recorded meteorological history. And somehow I, who almost never took that road, who rarely visited that part of the university, had found myself in its path.

Only much later did I realize that the tornado's eye had passed directly over me. It seemed to me that there was something eerily apt about that metaphor: what had happened at that moment was strangely like a species of visual contact, of beholding and being beheld. And in that instant of contact something was planted deep in my mind, something irreducibly mysterious, something quite apart from the danger that I had been in and the destruction that I had witnessed; something that was not a property of the thing itself but of the manner in which it had intersected with my life.

6.

As is often the case with people who are waylaid by unpredictable events, for years afterward my mind kept returning to my encounter with the tornado. Why had I walked down a road that I almost never took, just before it was struck by a phenomenon that was without historical precedent? To think of it in terms of chance and coincidence seemed only to impoverish the experience: it was like trying to understand a poem by counting the words. I found myself reaching instead for the opposite end of the spectrum of meaning – for the extraordinary, the inexplicable, the confounding. Yet these too did not do justice to my memory of the event.

Novelists inevitably mine their own experience when they write. Unusual events being necessarily limited in number, it is but natural that these should be excavated over and again, in the hope of discovering a yet undiscovered vein.

No less than any other writer have I dug into my own past while writing fiction. By rights then, my encounter with the tornado should have been a mother lode, a gift to be mined to the last little nugget.

It is certainly true that storms, floods, and unusual weather events do recur in my books, and this may well be a legacy of the tornado. Yet oddly enough, no tornado has ever figured in my novels. Nor is this due to any lack of effort on my part. Indeed, the reason I still possess those cuttings from the *Times of India* is that I have returned to them often over the years, hoping to put them to use in a novel, but only to meet with failure at every attempt.

On the face of it there is no reason why such an event should be difficult to translate into fiction; after all, many novels are filled with strange happenings. Why then did I fail, despite my best efforts, to send a character down a road that is imminently to be struck by a tornado?

In reflecting on this, I find myself asking, What would I make of such a scene were I to come across it in a novel written by someone else? I suspect that my response would be one of incredulity; I would be inclined to think that the scene was a contrivance of last resort. Surely only a writer whose imaginative resources were utterly depleted would fall back on a situation of such extreme improbability?

Improbability is the key word here, so we have to ask, What does the word mean?

Improbable is not the opposite of *probable*, but rather an inflexion of it, a gradient in a continuum of probability. But what does probability – a mathematical idea – have to do with fiction?

The answer is: Everything. For, as Ian Hacking, a prominent historian of the concept, puts it, probability is a 'manner of conceiving the world constituted without our being aware of it.'

Probability and the modern novel are in fact twins, born at about the same time, among the same people, under a shared star that destined them to work as vessels for the containment of the same kind of experience. Before the birth of the modern novel, wherever stories were told, fiction delighted in the unheard-of and the unlikely. Narratives like those of *The Arabian Nights*, *The Journey to the West*, and *The Decameron* proceed by leaping blithely from one exceptional event to another. This, after all, is how storytelling must necessarily proceed, inasmuch as it is a recounting of 'what happened' – for such an inquiry can arise only in relation to something out of the ordinary, which is but another way of saying 'exceptional' or 'unlikely.' In essence, narrative proceeds by linking together moments and scenes that are in some way distinctive or different: these are, of course, nothing other than instances of exception.

Novels too proceed in this fashion, but what is distinctive about the form is precisely the concealment of those exceptional moments that serve as the motor of narrative. This is achieved through the insertion of what Franco Moretti, the literary theorist, calls 'fillers.' According to Moretti, 'fillers function very much like the good manners so important in [Jane] Austen: they are both mechanisms designed to keep the "narrativity" of life under control – to give a regularity, a "style" to existence.' It is through this mechanism that worlds are conjured up, through everyday details, which function 'as *the opposite of narrative.*'

It is thus that the novel takes its modern form, through 'the relocation of the unheard-of toward the background . . . while the everyday moves into the foreground.'

Thus was the novel midwifed into existence around the world, through the banishing of the improbable and the insertion of the everyday. The process can be observed with exceptional clarity in the work of Bankim Chandra Chatterjee, a nineteenth-century Bengali writer and critic who self-consciously adopted the project of carving out a space in which realist European-style fiction could be written in the vernacular languages of India. Chatterjee's enterprise, undertaken in a context that was far removed from the metropolitan mainstream, is one of those instances in which a circumstance of exception reveals the true life of a regime of thought and practice.

Chatterjee was, in effect, seeking to supersede many old and very powerful forms of fiction, ranging from the ancient Indian epics to Buddhist Jataka stories and the immensely fecund Islamicate tradition of Urdu *dastaans*. Over time, these narrative forms had accumulated great weight and authority, extending far beyond the Indian subcontinent: his attempt to claim territory for a new kind of fiction was thus, in its own way, a heroic endeavor. That is why Chatterjee's explorations are of particular interest: his charting of this new territory puts the contrasts between the Western novel and other, older forms of narrative in ever-sharper relief.

In a long essay on Bengali literature, written in 1871, Chatterjee launched a frontal assault on writers who modeled their work on traditional forms of storytelling: his attack on this so-called Sanskrit school was focused precisely on the notion of 'mere narrative.' What he advocated instead was a style of writing that would accord primacy to 'sketches of character and pictures of Bengali life.'

What this meant, in practice, is very well illustrated by Chatterjee's first novel, *Rajmohan's Wife*, which was written in English in the early 1860s. Here is a passage: 'The house of Mathur Ghose was a genuine specimen of mofussil [provincial] magnificence united with a mofussil want of cleanliness . . . From the far-off paddy fields you could descry through the intervening foliage its high palisades and blackened walls. On a nearer view might be seen pieces of plaster of a venerable antiquity prepared to bid farewell to their old and weather-beaten tenement.'

Compare this with the following lines from Gustave Flaubert's *Madame Bovary*: 'We leave the high road . . .

whence the valley is seen ... The meadow stretches under a bulge of low hills to join at the back with the pasture land of the Bray country, while on the eastern side, the plain, gently rising, broadens out, showing as far as eye can follow its blond cornfields.'

In both these passages, the reader is led into a 'scene' through the eye and what it beholds: we are invited to 'descry,' to 'view,' to 'see.' In relation to other forms of narrative, this is indeed something new: instead of being told about what happened we learn about what was observed. Chatterjee has, in a sense, gone straight to the heart of the realist novel's 'mimetic ambition': detailed descriptions of everyday life (or 'fillers') are therefore central to his experiment with this new form.

Why should the rhetoric of the everyday appear at exactly the time when a regime of statistics, ruled by ideas of probability and improbability, was beginning to give new shapes to society? Why did fillers suddenly become so important? Moretti's answer is 'Because they *offer the kind of narrative pleasure compatible with the new regularity of bourgeois life*. Fillers turn the novel into a "calm passion" ... they are part of what Weber called the "rationalization" of modern life: a process that begins in the economy and in the administration, but eventually pervades the sphere of free time, private life, entertainment, feelings ... Or in other words: fillers are an attempt at rationalizing the novelistic universe: turning it into a world of few surprises, fewer adventures, and no miracles at all.'

This regime of thought imposed itself not only on the arts but also on the sciences. That is why *Time's Arrow*,

Time's Cycle, Stephen Jay Gould's brilliant study of the geological theories of gradualism and catastrophism is, in essence, a study of narrative. In Gould's telling of the story, the catastrophist recounting of the earth's history is exemplified by Thomas Burnet's *Sacred Theory of the Earth* (1690) in which the narrative turns on events of 'unrepeatable uniqueness.' As opposed to this, the gradualist approach, championed by James Hutton (1726–97) and Charles Lyell (1797–1875), privileges slow processes that unfold over time at even, predictable rates. The central credo in this doctrine was 'nothing could change otherwise than the way things were seen to change in the present.' Or, to put it simply: 'Nature does not make leaps.'

The trouble, however, is that Nature does certainly jump, if not leap. The geological record bears witness to many fractures in time, some of which led to mass extinctions and the like: it was one such, in the form of the Chicxulub asteroid, that probably killed the dinosaurs. It is indisputable, in any event, that catastrophes waylay both the earth and its individual inhabitants at unpredictable intervals and in the most improbable ways.

Which, then, has primacy in the real world, predictable processes or unlikely events? Gould's response is 'the only possible answer can be "both and neither." ' Or, as the National Research Council of the United States puts it: 'It is not known whether the relocation of materials on the surface of the Earth is dominated by the slower but continuous fluxes operating all the time or by the spectacular large fluxes that operate during short-lived cataclysmic events.'

It was not until quite recently that geology reached this agnostic consensus. Through much of the era when geology – and also the modern novel – were coming of age, the gradualist (or 'uniformitarian') view held absolute sway and catastrophism was exiled to the margins. Gradualists consolidated their victory by using one of modernity's most effective weapons: its insistence that it has rendered other forms of knowledge obsolete. So, as Gould so beautifully demonstrates, Lyell triumphed over his adversaries by accusing them of being primitive: 'In an early stage of advancement, when a great number of natural appearances are unintelligible, an eclipse, an earthquake, a flood, or the approach of a comet, with many other occurrences afterwards found to belong to the regular course of events, are regarded as prodigies. The same delusion prevails as to moral phenomena, and many of these are ascribed to the intervention of demons, ghosts, witches, and other immaterial and supernatural agents.'

This is exactly the rhetoric that Chatterjee uses in attacking the 'Sanskrit school': he accuses those writers of depending on conventional modes of expression and fantastical forms of causality. 'If love is to be the theme, Madana is invariably put into requisition with his five flower-tipped arrows; and the tyrannical king of Spring never fails to come to fight in his cause, with his army of bees, and soft breezes, and other ancient accompaniments. Are the pangs of separation to be sung? The moon is immediately cursed and anathematized, as scorching the poor victim with her cold beams.'

Flaubert sounds a strikingly similar note in satirizing

the narrative style that entrances the young Emma Rou-
ault: in the novels that were smuggled into her convent,
it was 'all love, lovers, sweethearts, persecuted ladies
fainting in lonely pavilions, postilions killed at every
stage, horses ridden to death on every page, sombre forests, heartaches, vows, sobs, tears and kisses, little skiffs
by moonlight, nightingales in shady groves.' All of this is
utterly foreign to the orderly bourgeois world that Emma
Bovary is consigned to; such fantastical stuff belongs in
the 'dithyrambic lands' that she longs to inhabit.

In a striking summation of her tastes in narrative,
Emma declares, 'I . . . adore stories that rush breath-
lessly along, that frighten one. I detest commonplace
heroes and moderate sentiments, such as there are in
Nature.'

'Commonplace'? 'Moderate'? How did Nature ever
come to be associated with words like these?

The incredulity that these associations evoke today is
a sign of the degree to which the Anthropocene has
already disrupted many assumptions that were founded
on the relative climatic stability of the Holocene. From
the reversed perspective of our time, the complacency
and confidence of the emergent bourgeois order appears
as yet another of those uncanny instances in which the
planet seems to have been toying with humanity, by
allowing it to assume that it was free to shape its own
destiny.

Unlikely though it may seem today, the nineteenth
century was indeed a time when it was assumed, in
both fiction and geology, that Nature was moderate
and orderly: this was a distinctive mark of a new and

'modern' worldview. Chatterjee goes out of his way to berate his contemporary, the poet Michael Madhusu-dan Datta, for his immoderate portrayals of Nature: 'Mr. Datta . . . wants repose. The winds rage their loud-est when there is no necessity for the lightest puff. Clouds gather and pour down a deluge, when they need do nothing of the kind; and the sea grows terrible in its wrath, when everybody feels inclined to resent its interference.'

The victory of gradualist views in science was similarly won by characterizing catastrophism as un-modern. In geology, the triumph of gradualist thinking was so com-plete that Alfred Wegener's theory of continental drift, which posited upheavals of sudden and unimaginable violence, was for decades discounted and derided.

It is worth recalling that these habits of mind held sway until late in the twentieth century, especially among the general public. 'As of the mid-1960s,' writes the historian John L. Brooke, 'a gradualist model of earth history and evolution . . . reigned supreme.' Even as late as 1985, the editorial page of the *New York Times* was inveighing against the asteroidal theory of dinosaur extinction: 'Astronomers should leave to astrologers the task of seeking the causes of events in the stars.' As for professional paleontologists, Elizabeth Kolbert notes, they reviled both the theory and its originators, Luis and Walter Alvarez: ' "The Cretaceous extinctions were grad-ual and the catastrophe theory is wrong," . . . [a] paleontologist stated. But "simplistic theories will con-tinue to come along to seduce a few scientists and enliven the covers of popular magazines." '

In other words, gradualism became 'a set of blinders' that eventually had to be put aside in favor of a view that recognizes the 'twin requirements of uniqueness to mark moments of time as distinctive, and lawfulness to establish a basis of intelligibility.'

Distinctive moments are no less important to modern novels than they are to any other forms of narrative, whether geological or historical. Ironically, this is no-where more apparent than in *Rajmohan's Wife* and *Madame Bovary*, in both of which chance and happen-stance are crucial to the narrative. In Flaubert's novel, for instance, the narrative pivots at a moment when Mon-sieur Bovary has an accidental encounter with his wife's soon-to-be lover at the opera, just after an impassioned scene during which she has imagined that the lead singer 'was looking at her . . . She longed to run to his arms, to take refuge in his strength, as in the incarnation of love itself, and to say to him, to cry out, "Take me away! carry me with you!"'

It could not, of course, be otherwise: if novels were not built upon a scaffolding of exceptional moments, writers would be faced with the Borgesian task of repro-ducing the world in its entirety. But the modern novel, unlike geology, has never been forced to confront the centrality of the improbable: the concealment of its scaffolding of events continues to be essential to its func-tioning. It is this that makes a certain kind of narrative a recognizably modern novel.

Here, then, is the irony of the 'realist' novel: the very gestures with which it conjures up reality are actually a concealment of the real.

What this means in practice is that the calculus of probability that is deployed within the imaginary world of a novel is not the same as that which obtains outside it; this is why it is commonly said, 'If this were in a novel, no one would believe it.' Within the pages of a novel an event that is only slightly improbable in real life – say, an unexpected encounter with a long-lost childhood friend – may seem wildly unlikely: the writer will have to work hard to make it appear persuasive.

If that is true of a small fluke of chance, consider how much harder a writer would have to work to set up a scene that is wildly improbable even in real life? For example, a scene in which a character is walking down a road at the precise moment when it is hit by an unheard-of weather phenomenon?

To introduce such happenings into a novel is in fact to court eviction from the mansion in which serious fiction has long been in residence; it is to risk banishment to the humbler dwellings that surround the manor house – those generic outhouses that were once known by names such as 'the Gothic,' 'the romance,' or 'the melodrama,' and have now come to be called 'fantasy,' 'horror,' and 'science fiction.'

7.

So far as I know, climate change was not a factor in the tornado that struck Delhi in 1978. The only thing it has in common with the freakish weather events of today is its extreme improbability. And it appears that we are now in an era that will be defined precisely by events that appear, by our current standards of normalcy, highly improbable: flash floods, hundred-year storms, persistent droughts, spells of unprecedented heat, sudden landslides, raging torrents pouring down from breached glacial lakes, and, yes, freakish tornadoes.

The superstorm that struck New York in 2012, Hurricane Sandy, was one such highly improbable phenomenon: the word *unprecedented* has perhaps never figured so often in the description of a weather event. In his fine study of Hurricane Sandy, the meteorologist Adam Sobel notes that the track of the storm, as it crashed into the east coast of the United States, was without precedent: never before had a hurricane veered sharply westward in the mid-Atlantic. In turning, it also merged with a winter storm, thereby becoming a 'mammoth hybrid' and attaining a size unprecedented in scientific memory. The storm surge that it unleashed reached a height that exceeded any in the region's recorded meteorological history.

Indeed, Sandy was an event of such a high degree of improbability that it confounded statistical weather-prediction models. Yet dynamic models, based on the laws of physics, were able to accurately predict its trajectory as well as its impacts.

But calculations of risk, on which officials base their decisions in emergencies, are based largely on probabilities. In the case of Sandy, as Sobel shows, the essential improbability of the phenomenon led them to underestimate the threat and thus delay emergency measures.

Sobel goes on to make the argument, as have many

others, that human beings are intrinsically unable to prepare for rare events. But has this really been the case throughout human history? Or is it rather an aspect of the unconscious patterns of thought – or 'common sense' – that gained ascendancy with a growing faith in 'the regularity of bourgeois life'? I suspect that human beings were generally catastrophists at heart until their instinctive awareness of the earth's unpredictability was gradually supplanted by a belief in uniformitarianism – a regime of ideas that was supported by scientific theories like Lyell's, and also by a range of governmental practices that were informed by statistics and probability.

It is a fact, in any case, that when early tremors jolted the Italian town of L'Aquila, shortly before the great earthquake of 2009, many townsfolk obeyed the instinct that prompts people who live in earthquake-prone areas to move to open spaces. It was only because of a governmental intervention, intended to prevent panic, that they returned to their homes. As a result, a good number were trapped indoors when the earthquake occurred.

No such instinct was at work in New York during Sandy, where, as Sobel notes, it was generally believed that 'losing one's life to a hurricane is . . . something that happens in faraway places' (he might just as well have said 'dithyrambic lands'). In Brazil, similarly, when Hurricane Catarina struck the coast in 2004, many people did not take shelter because 'they refused to believe that hurricanes were possible in Brazil.'

But in the era of global warming, nothing is really far away; there is no place where the orderly expectations of bourgeois life hold unchallenged sway. It is as though our earth had become a literary critic and were laughing at Flaubert, Chatterjee, and their like, mocking *their* mockery of the 'prodigious happenings' that occur so often in romances and epic poems.

This, then, is the first of the many ways in which the age of global warming defies both literary fiction and contemporary common sense: the weather events of this time have a very high degree of improbability. Indeed, it has even been proposed that this era should be named the 'catastrophozoic' (others prefer such phrases as 'the long emergency' and 'the Penumbral Period'). It is certain in any case that these are not ordinary times: the events that mark them are not easily accommodated in the deliberately prosaic world of serious prose fiction.

Poetry, on the other hand, has long had an intimate relationship with climatic events: as Geoffrey Parker points out, John Milton began to compose *Paradise Lost* during a winter of extreme cold, and 'unpredictable and unforgiving changes in the climate are central to his story. Milton's fictional world, like the real one in which he lived, was . . . a "universe of death" at the mercy of extremes of heat and cold.' This is a universe very different from that of the contemporary literary novel.

I am, of course, painting with a very broad brush: the novel's infancy is long past, and the form has changed in many ways over the last two centuries. Yet, to a quite remarkable degree, the literary novel has also remained true to the destiny that was charted for it at birth. Consider that the literary movements of the twentieth century were almost uniformly disdainful of plot and narrative; that an ever-greater emphasis was laid on style and 'observation,' whether it be of everyday details, traits of character, or nuances of emotion – which is why teachers of creative writing now exhort their students to 'show, don't tell.'

Yet fortunately, from time to time, there have also been movements that celebrated the unheard-of and the improbable: surrealism for instance, and most significantly, magical realism, which is replete with events that have no relation to the calculus of probability.

There is, however, an important difference between the weather events that we are now experiencing and those that occur in surrealist and magical realist novels: improbable though they might be, these events are neither surreal nor magical. To the contrary, these highly improbable occurrences are overwhelmingly, urgently, astoundingly real. The ethical difficulties that might arise in treating them as magical or metaphorical or allegorical are obvious perhaps. But there is another reason why, from the writer's point of view, it would serve no purpose to approach them in that way: because to treat them as magical or surreal would be to rob them of precisely the quality that makes them so urgently compelling – which is that they are actually happening on this earth, at this time.

8.

The Sundarbans are nothing like the forests that usually figure in literature. The greenery is dense, tangled, and low; the canopy is not above but around you, constantly clawing at your skin and your clothes. No breeze can enter the thickets of this forest; when the air stirs at all it is because of the buzzing of flies and other insects. Underfoot, instead of a carpet of softly decaying foliage, there is a bank of slippery, knee-deep mud, perforated by the sharp points that protrude from mangrove roots. Nor do any vistas present themselves except when you are on one of the hundreds of creeks and channels that wind through the landscape – and even then it is the water alone that opens itself; the forest withdraws behind its muddy ramparts, disclosing nothing.

In the Sundarbans, tigers are everywhere and nowhere. Often when you go ashore, you will find fresh tiger prints in the mud, but of the animal itself you will see nothing: glimpses of tigers are exceedingly uncommon and rarely more than fleeting. Yet you cannot doubt, since the prints are so fresh, that a tiger is somewhere nearby; and you know that it is probably watching you. In this jungle, concealment is so easy for an animal that it could be just a few feet away. If it charged, you would not see it till the last minute, and even if you did, you would not be able to get away; the mud would immobilize you.

Scattered through the forest are red rags, fluttering on branches. These mark the sites where people have been killed by tigers. There are many such killings every year; exactly how many no one knows because the statistics are not reliable. Nor is this anything new; in the nineteenth century, tens of thousands were killed by tigers. Suffice it to say that in some villages every household has lost a member to a tiger; everyone has a story to tell.

In these stories a great deal hinges on the eyes; seeing is one of their central themes; *not* seeing is another. The tiger is watching you; you are aware of its gaze, as you always are, but you do not see it; you do not lock eyes with it until it launches its charge, and at that moment a shock courses through you and you are immobilized, frozen.

The folk epic of the Sundarbans, *Bon Bibir Johuranama (The Miracles of Bon Bibi)*, comes to a climax in one such moment of mutual beholding, when the tiger demon, Dokkhin Rai, locks eyes with the protagonist, a boy called Dukhey:

> It was then from afar, that the demon saw Dukhey . . .
>
> Long had he hungered for this much-awaited prize; in an instant he assumed his tiger disguise.
>
> 'How long has it been since human flesh came my way? Now bliss awaits me in the shape of this boy Dukhey.'
>
> On the far mudbank Dukhey caught sight of the beast: 'that tiger is the demon and I'm to be his feast.'
>
> Raising its head, the tiger reared its immense back; its jowls filled like sails as it sprang to attack.
>
> The boy's life took wing, on seeing this fearsome sight.

Many stories of encounters with tigers hinge upon a moment of mutual recognition like this one. To look into the tiger's eyes is to recognize a presence of which you are already aware; and in that moment of contact you realize that this presence possesses a similar awareness of you, even though it is not human. This mute exchange of gazes is the only communication that is possible between you and this presence – yet communication it undoubtedly is.

But what is it that you are communicating with, at this moment of extreme danger, when your mind is in a state unlike any you've ever known before? An analogy that is sometimes offered is that of seeing a ghost, a presence that is not of this world.

In the tiger stories of the Sundarbans, as in my experience of the tornado, there is, as I noted earlier, an irreducible element of mystery. But what I am trying to suggest is perhaps better expressed by a different word, one that recurs frequently in translations of Freud and Heidegger. That word is *uncanny* – and it is indeed with uncanny accuracy that my experience of the tornado is evoked in the following passage: 'In dread, as we say, "one feels something uncanny." What is this "something" and this "one"? We are unable to say what gives "one" that uncanny feeling. "One" just feels it generally.'

It is surely no coincidence that the word *uncanny* has begun to be used, with ever greater frequency, in relation to climate change. Writing of the freakish events and objects of our era, Timothy Morton asks, 'Isn't it the case, that the effect delivered to us in the [unaccustomed] rain, the weird cyclone, the oil slick is something uncanny?' George Marshall writes, 'Climate change is inherently uncanny: Weather conditions, and the high-carbon lifestyles that are changing them, are extremely familiar and yet have now been given a new menace and uncertainty.'

No other word comes close to expressing the strangeness of what is unfolding around us. For these changes are not merely strange in the sense of being unknown or alien; their uncanniness lies precisely in the fact that in these encounters we recognize something we had turned away from: that is to say, the presence and proximity of nonhuman interlocutors.

Yet now our gaze seems to be turning again; the uncanny and improbable events that are beating at our doors seem to have stirred a sense of recognition, an awareness that humans were never alone, that we have always been surrounded by beings of all sorts who share elements of that which we had thought to be most distinctively our own: the capacities of will, thought, and consciousness. How else do we account for the interest in the nonhuman that has been burgeoning in the humanities over the last decade and over a range of disciplines; how else do we account for the renewed attention to panpsychism and the metaphysics of Alfred North Whitehead; and for the rise to prominence of object-oriented ontology, actor–network theory, the new animism, and so on?

Can the timing of this renewed recognition be mere coincidence, or is the synchronicity an indication that there are entities in the world, like forests, that are fully capable of inserting themselves into our processes of thought? And if that were so, could it not *also* be said that the earth has itself intervened to revise those habits of thought that are based on the Cartesian dualism that arrogates all intelligence and agency to the human while denying them to every other kind of being?

This possibility is not, by any means, the most important of the many ways in which climate change challenges and refutes Enlightenment ideas. It is, however, certainly the most uncanny. For what it suggests – indeed proves – is that nonhuman forces have the ability to intervene directly in human thought. And to be alerted to such interventions is also to become uncannily aware that conversations among ourselves have always had other participants: it is like finding out that one's telephone has been tapped for years, or that the neighbors have long been eavesdropping on family discussions.

But in a way it's worse still, for it would seem that those unseen presences actually played a part in shaping our discussions without our being aware of it. And if these are real possibilities, can we help but suspect that all the time that we imagined ourselves to be thinking about apparently inanimate objects, we were ourselves being 'thought' by other entities? It is almost as if the mind-altering planet that Stanislaw Lem imagined in *Solaris* were our own, familiar Earth: what could be more uncanny than this?

These possibilities have many implications for the subject that primarily concerns me here, literary fiction. I will touch on some of these later, but for now I want to attend only to the aspect of the uncanny.

On the face of it, the novel as a form would seem to be a natural home for the uncanny. After all, have not some of the greatest novelists written uncanny tales? The ghost stories of Charles Dickens, Henry James, and Rabindranath Tagore come immediately to mind.

But the environmental uncanny is not the same as the

uncanniness of the supernatural: it is different precisely because it pertains to nonhuman forces and beings. The ghosts of literary fiction are not human either, of course, but they are certainly represented as projections of humans who were once alive. But animals like the Sundarbans tiger, and freakish weather events like the Delhi tornado, have no human referents at all.

There is an additional element of the uncanny in events triggered by climate change, one that did not figure in my experience of the Delhi tornado. This is that the freakish weather events of today, despite their radically nonhuman nature, are nonetheless animated by cumulative human actions. In that sense, the events set in motion by global warming have a more intimate connection with humans than did the climatic phenomena of the past – this is because we have all contributed in some measure, great or small, to their making. They are the mysterious work of our own hands returning to haunt us in unthinkable shapes and forms.

All of this makes climate change events peculiarly resistant to the customary frames that literature has applied to 'Nature': they are too powerful, too grotesque, too dangerous, and too accusatory to be written about in a lyrical, or elegiac, or romantic vein. Indeed, in that these events are not entirely of Nature (whatever that might be), they confound the very idea of 'Nature writing' or ecological writing: they are instances, rather, of the uncanny intimacy of our relationship with the nonhuman.

More than a quarter century has passed since Bill McKibben wrote, 'We live in a post-natural world.' But did 'Nature' in this sense ever exist? Or was it rather the deification of the human that gave it an illusory apartness from ourselves? Now that nonhuman agencies have dispelled that illusion, we are confronted suddenly with a new task: that of finding other ways in which to imagine the unthinkable beings and events of this era.

9.

In the final part of my novel *The Hungry Tide*, there is a scene in which a cyclone sends a gigantic storm surge into the Sundarbans. The wave results in the death of one of the principal characters, who gives his life protecting another.

This scene was extraordinarily difficult to write. In preparation for it, I combed through a great deal of material on catastrophic waves – storm surges as well as tsunamis. In the process, as often happens in writing fiction, the plight of the book's characters, as they faced the wave, became frighteningly real.

The Hungry Tide was published in the summer of 2004. A few months after the publication, on the night of December 25, I was back in my family home in Kolkata. The next morning, on logging on to the web, I learned that a cataclysmic tsunami had been set off by a massive undersea earthquake in the Indian Ocean. Measuring 9.0 on the Richter scale, the quake's epicenter lay between the northernmost tip of Sumatra and the southernmost island in the Andaman and Nicobar chain. Although the full extent of the catastrophe was not yet known, it was already clear that the toll in human lives would be immense.

The news had a deeply unsettling effect on me: the images that had been implanted in my mind by the writing of *The Hungry Tide* merged with live television footage of the tsunami in a way that was almost overwhelming. I became frantic; I could not focus on anything.

A couple of days later, I wrote to a newspaper and obtained a commission to write about the impact of the tsunami on the Andaman and Nicobar Islands. My first stop was the islands' capital city, Port Blair, which was thronged with refugees but had not suffered much damage itself: its location, above a sheltered cove, had protected it. After spending a few days there, I was able to board an Indian Air Force plane that was carrying supplies to one of the worst affected of the Nicobar Islands.

Unlike the Andamans, which rise steeply from the sea, the Nicobars are low-lying islands. Being situated close to the quake's epicenter, they had been very badly hit; many settlements had been razed. I visited a shore-side town called Malacca that had been reduced literally to its foundations: of the houses only the floors were left, and here and there the stump of a wall. It was as though the place had been hit by a bomb that was designed specifically to destroy all things human – for one of the strangest aspects of the scene was that the island's coconut palms were largely unaffected; they stood serenely amid the rubble, their fronds waving gently in the breeze that was blowing in from the sparkling, sun-drenched sea.

I wrote in my notebook: 'The damage was limited to

a half-mile radius along the shore. In the island's interior everything is tranquil, peaceful – indeed astonishingly beautiful. There are patches of tall, dark primary forest, beautiful padauk trees, and among these, in little clearings, huts built on stilts . . . One of the ironies of the situation is that the most upwardly mobile people on the island were living at its edges.'

Such was the pattern of settlement here that the indigenous islanders lived mainly in the interior: they were largely unaffected by the tsunami. Those who had settled along the seashore, on the other hand, were mainly people from the mainland, many of whom were educated and middle class: in settling where they had, they had silently expressed their belief that highly improbable events belong not in the real world but in fantasy. In other words, even here, in a place about as far removed as possible from the metropolitan centers that have shaped middle-class lifestyles, the pattern of settlement had come to reflect the uniformitarian expectations that are rooted in the 'regularity of bourgeois life.'

At the air force base where my plane had landed there was another, even more dramatic, illustration of this. The functional parts of the base – where the planes and machinery were kept – were located to the rear, well away from the water. The living areas, comprised of pretty little two-story houses, were built much closer to the sea, at the edge of a beautiful, palm-fringed beach. As always in military matters, the protocols of rank were strictly observed: the higher the rank of the officers, the closer their houses were to the water and the better the view that they and their families enjoyed.

Such was the design of the base that when the tsunami struck these houses the likelihood of survival was small, and inasmuch as it existed at all, it was in inverse relation to rank: the commander's house was thus the first to be hit.

The sight of the devastated houses was disturbing for reasons beyond the immediate tragedy of the tsunami and the many lives that had been lost there: the design of the base suggested a complacency that was itself a kind of madness. Nor could the siting of these buildings be attributed to the usual improvisatory muddle of Indian patterns of settlement. The base had to have been designed and built by a government agency; the site had clearly been chosen and approved by hardheaded military men and state-appointed engineers. It was as if, in being adopted by the state, the bourgeois belief in the regularity of the world had been carried to the point of derangement.

A special place ought to be reserved in hell, I thought to myself, for planners who build with such reckless disregard for their surroundings.

Not long afterward, while flying into New York's John F. Kennedy airport, I looked out of the window and spotted Far Rockaway and Long Beach, the thickly populated Long Island neighborhoods that separate the airport from the Atlantic Ocean. Looking down on them from above, it was clear that those long rows of apartment blocks were sitting upon what had once been barrier islands, and that in the event of a major storm surge they would be swamped (as indeed they were when Hurricane Sandy hit the area in 2012). Yet it was clear also that these neighborhoods had not sprung up haphazardly; the sanction of the state was evident in the ordered geometry of their streets.

Only then did it strike me that the location of that base in the Nicobars was by no means anomalous; the builders had not in any sense departed from accepted global norms. To the contrary, they had merely followed the example of the European colonists who had founded cities like Bombay (Mumbai), Madras (Chennai), New York, Singapore, and Hong Kong, all of which are sited directly on the ocean. I understood also that what I had seen in the Nicobars was but a microcosmic expression of a pattern of settlement that is now dominant around the world: proximity to the water is a sign of affluence and education; a beachfront location is a status symbol; an ocean view greatly increases the value of real estate. A colonial vision of the world, in which proximity to the water represents power and security, mastery and conquest, has now been incorporated into the very foundations of middle-class patterns of living across the globe.

But haven't people always liked to live by the water?

Not really; through much of human history, people regarded the ocean with great wariness. Even when they made their living from the sea, through fishing or trade, they generally did not build large settlements on the water's edge: the great old port cities of Europe, like London, Amsterdam, Rotterdam, Stockholm, Lisbon, and Hamburg, are all protected from the open ocean by bays, estuaries, or deltaic river systems. The same is true of old Asian ports: Cochin, Surat, Tamluk, Dhaka,

Mrauk-U, Guangzhou, Hangzhou, and Malacca are all cases in point. It is as if, before the early modern era, there had existed a general acceptance that provision had to be made for the unpredictable furies of the ocean – tsunamis, storm surges, and the like.

An element of that caution seems to have lingered even after the age of European global expansion began in the sixteenth century: it was not till the seventeenth century that colonial cities began to rise on seafronts around the world. Mumbai, Chennai, New York, and Charleston were all founded in this period. This would be followed by another, even more confident wave of city building in the nineteenth century, with the founding of Singapore and Hong Kong. These cities, all brought into being by processes of colonization, are now among those that are most directly threatened by climate change.

10.

Mumbai and New York, so different in so many ways, have in common that their destinies came to be linked to the British Empire at about the same time: the 1660s.

Although Giovanni da Verrazzano landed on Manhattan in 1524, the earliest European settlements in what is now New York State were built a long way up the Hudson River, in the area around Albany. It was not till 1625 that the Dutch built Fort Amsterdam on Manhattan island; this would later become New Amsterdam and then, when the British first seized the settlement in the 1660s, New York.

The site of today's Mumbai first came under European rule in 1535 when it was ceded to the Portuguese by the ruler of Gujarat. The site consisted of an estuarine archipelago, with a couple of large hilly islands to the north, close to the mainland, and a cluster of mainly low-lying islands to the south. This being an estuarine region, the relationship between land and water was so porous that the topography of the archipelago varied with the tides and the seasons.

Networks of shrines, villages, forts, harbors, and bazaars had existed on the southern islands for millennia, but they were never the site of an urban center as such, even in the early years of European occupation. The Portuguese built several churches and fortifications on those islands, but their main settlements were located close to the mainland, at Bassein, and on Salsette.

The southern part of the archipelago passed into British control when King Charles II married Catherine of Braganza in 1661: the islands were included in her dowry (which also contained a chest of tea: this was the Pandora's box that introduced the British public to the beverage, thereby setting in motion the vast cycles of trade that would turn nineteenth-century Bombay into the world's leading opium exporting port). It was only after passing into British hands that the southern islands became the nucleus of a sprawling urban conglomeration. It was then too that a distinct line of separation between land and sea was conjured up through the application of techniques of surveying within a 'milieu of colonial power.'

The appeal of the sites of both Mumbai and New York

lay partly in their proximity to deepwater harbors and partly in the strategic advantages they presented: as islands, they were both easier to defend and easier to supply from the metropolis. A certain precariousness was thus etched upon them from the start by reason of their colonial origins.

The islands of south Mumbai did not long remain as they were when they were handed over to the British: links between them, in the form of causeways, bridges, embankments, and reclamation projects, began to rise in the eighteenth century. The reshaping of the estuarine landscape proceeded at such a pace that by the 1860s a Marathi chronicler, Govind Narayan, was able to predict with confidence that soon it would 'never occur to anybody that Mumbai was an island once.'

Today the part of the city that is located on the former islands to the south of Salsette has a population of about 11.8 million (the population of the Greater Mumbai area is somewhere in the region of 19 to 20 million). This promontory, less than twenty kilometers in length, is the center of many industries, including India's financial industry; the adjoining port handles more than half of the country's containerized cargo. This part of Mumbai is also home to many millionaires and billionaires: naturally many of them live along the western edge of the peninsula, which offers the finest views of the Arabian Sea.

Because of the density of its population and the importance of its institutions and industries, Mumbai represents an extraordinary, possibly unique, 'concentration of risk.' For this teeming metropolis, this great hub of economic, financial, and cultural activity, sits upon a wedge of cobbled-together land that is totally exposed to the ocean. It takes only a glance at a map to be aware of this: yet it was not till 2012 when Superstorm Sandy barreled into New York on October 29 that I began to think about the dangers of Mumbai's topography.

My wife and I were actually in Goa at the time, but since New York is also home to us we followed the storm closely, on the web and on TV, watching with mounting apprehension and disbelief as the storm swept over the city, devastating the oceanfront neighborhoods that we had flown over so many times while coming in to land at JFK airport.

As I watched these events unfold it occurred to me to wonder what would happen if a similar storm were to hit Mumbai. I reassured myself with the thought that this was very unlikely: both Mumbai and Goa face the Arabian Sea, which, unlike the Bay of Bengal, has not historically generated a great deal of cyclonic activity. Nor, unlike India's east coast, has the west coast had to deal with tsunamis: it was unaffected by the tsunami of 2004, for instance, which devastated large stretches of the eastern seaboard.

Still, the question intrigued me and I began to hunt for more information on the region's seismic and cyclonic profiles. Soon enough I learned that the west coast's good fortune might be merely a function of the providential protraction of geological time – for the Arabian Sea is by no means seismically inactive. A previously unknown, and probably very active, fault system was discovered in the Owen fracture zone a few years ago, off the coast of Oman; the system is eight hundred kilometers long and faces the west coast of India. This discovery was announced in an article that concludes with these words, chilling in their understatement: 'These results will motivate a reappraisal of the seismic and tsunami hazard assessment in the NW Indian Ocean.'

Soon, I also had to rethink my assumptions about cyclones and the Arabian Sea. Reading about Hurricane Sandy, I came upon more and more evidence that climate change may indeed alter patterns of cyclonic activity around the world: Adam Sobel's *Storm Surge*, for example, suggests that significant changes may be in the offing. When I began to look for information on the Arabian Sea in particular, I learned that there had been an uptick in cyclonic activity in those waters over the last couple of decades. Between 1998 and 2001, three cyclones had crashed into the Indian subcontinent to the north of Mumbai: they claimed over seventeen thousand lives. Then in 2007, the Arabian Sea generated its strongest ever recorded storm: Cyclone Gonu, a Category 5 hurricane, which hit Oman, Iran, and Pakistan in June that year causing widespread damage.

What do these storms portend? Hoping to find an answer, I reached out to Adam Sobel, who is a professor of atmospheric science at Columbia University. He agreed to an interview, and on a fine October day in 2015, I made my way to his Manhattan apartment. He confirmed to me that the most up-to-date research indicates that the Arabian Sea is one of the regions of the world where cyclonic activity is indeed likely to increase: a 2012 paper by a Japanese research team predicts a 46 percent increase in tropical cyclone frequency in the Arabian Sea by the end of the next century, with a corresponding 31 percent decrease in the Bay of Bengal. It also predicts another change: in the past, cyclones were rare during the monsoon because wind flows in the northern Indian Ocean were not conducive to their formation in that season. Those patterns are now changing in such a way as to make cyclones more likely during and after the monsoons. Another paper, by an American research team, concludes that cyclonic activity in the Arabian Sea is also likely to intensify because of the cloud of dust and pollution that now hangs over the Indian subcontinent and its surrounding waters: this too is contributing to changes in the region's wind patterns.

These findings prompted me to ask Adam whether he might be willing to write a short piece assessing the risks that changing climatic patterns pose for Mumbai. He agreed and thus began a very interesting series of exchanges.

A few weeks after our meeting, Adam sent me this message:

I have been doing a little research on Mumbai storm surge risk. There seems to be very little written about it. I have found a number of vague acknowledgments that the risk exists, but nothing that quantifies it.

However, are you aware of the 1882 Mumbai cyclone? I have found only very brief accounts of it so far, but the death toll appears to have been between 100,000 and 200,000, and one source says there was a 6m storm surge, which is enormous, and I presume would account for much if not all of that! This was in one paragraph of a book that seems to be out of print. I haven't quickly found any more substantive sources online – most are single-line mentions in lists of deadly storms. I wonder if you have ever seen anything more in-depth?

It is very spooky indeed that this storm is not mentioned in the various academic studies I have dug up on storm surge risk in India.

A quick Google search produced a number of references to an 1882 Bombay cyclone (some were even accompanied by pictures). There were several mentions of a death toll upward of one hundred thousand.

The figure astounded me. Mumbai's population then was about eight hundred thousand, which would mean that an eighth or more of the population would have perished: an extrapolation from these figures to today's Mumbai would yield a number of over a million.

But then came a surprise: Adam wrote to say that the 1882 cyclone was probably a hoax or rumor. He had not been able to find a reliable record of it; nor had any of the meteorologists or historians that he had written to. I then wrote to Murali Ranganathan, an expert on nineteenth-century Bombay, and he looked up the 1882 issues of the *Kaiser-i-Hind*, a Bombay-based Gujarati weekly run by Parsis. He found a brief description of a storm with strong winds and heavy rain on June 4, 1882, but there was no mention of any loss of life. Evidently, there was no great storm in 1882: it is a myth that has gained a life of its own.

However, the search did confirm that colonial Bombay had been struck by cyclones several times in the past;

the 1909 edition of the city's *Gazetteer* even notes, 'Since written history supplanted legend Bombay appears to have been visited somewhat frequently by great hurricanes and minor cyclonic storms.'

Mumbai's earliest recorded encounter with a powerful storm was on May 15, 1618. A Jesuit historian described it thus: 'The sky clouded, thunder burst, and a mighty wind arose. Towards nightfall a whirlwind raised the waves so high that the people, half dead from fear, thought that their city would be swallowed up . . . The whole was like the ruin at the end of all things.' Another Portuguese historian noted of this storm: 'The sea was brought into the city by the wind; the waves roared fearfully; the tops of the churches were blown off and immense stones were driven to vast distances; two thousand persons were killed.' If this figure is correct, it would suggest that the storm killed about a fifth of the population that then lived on the archipelago.

In 1740, another 'terrific storm' caused great damage to the city, and in 1783 a storm that was 'fatal to every ship in its path' killed four hundred people in Bombay harbor. The city was also hit by several cyclones in the nineteenth century: the worst was in 1854, when 'property valued at half-a-million pounds sterling' was destroyed in four hours and a thousand people were killed.

Since the late nineteenth century onward, cyclones in the region seem to have 'abated in number and intensity,' but that may well be changing now. In 2009 Mumbai did experience a cyclonic storm, but fortunately its maximum wind speeds were in the region of 50 mph (85 kmph), well below those of a Category 1 hurricane on the Saffir–Simpson hurricane intensity scale. But encounters with storms of greater intensity may be forthcoming: 2015 was the first year in which the Arabian Sea is known to have generated more storms than the Bay of Bengal. This trend could tip the odds toward the recurrence of storms like those of centuries past.

Indeed, even as Adam and I were exchanging messages, Cyclone Chapala, a powerful storm, was forming in the Arabian Sea. Moving westward, it would hit the coast of Yemen on November 3, becoming the first Category 1 cyclone in recorded history to do so: in just two days, it would deluge the coast with more rain than it would normally get in several years. And then – as if to confirm the projections – even as Chapala was still battering Yemen, another cyclone, Megh, formed in the Arabian Sea and began to move along a similar track. A few days later another cyclone began to take shape in the Bay of Bengal, so that the Indian subcontinent was flanked by cyclones on both sides, a very rare event.

Suddenly the waters around India were churning with improbable events.

II.

What might happen if a Category 4 or 5 storm, with 150 mph or higher wind speeds, were to run directly into Mumbai? Mumbai's previous encounters with powerful cyclones occurred at a time when the city had considerably less than a million inhabitants; today it is the second-largest municipality in the world with a population of over 20 million. With the growth of the city, its built environment has also changed so that weather that is by no means exceptional often has severe effects: monsoon downpours, for instance, often lead to flooding nowadays. With an exceptional event the results can be catastrophic.

One such occurred on July 26, 2005, when a downpour without precedent in Mumbai's recorded history descended on the city: the northern suburbs received 94.4 cm of rain in fourteen hours, one of the highest rainfall totals ever recorded anywhere in a single day. On that day, with catastrophic suddenness, the people of the city were confronted with the costs of three centuries of interference with the ecology of an estuarine location.

The remaking of the landscape has so profoundly changed the area's topography that its natural drainage channels are now little more than filth-clogged ditches. The old waterways have been so extensively filled in, diverted, and built over that their carrying capacity has been severely diminished; and the water bodies, swamplands, and mangroves that might have served as natural sinks have also been encroached upon to a point where they have lost much of their absorptive ability.

A downpour as extreme as that of July 26 would pose a challenge even to a very effective drainage system: Mumbai's choked creeks and rivers were wholly inadequate to the onslaught. They quickly overflowed causing floods in which water was mixed with huge quantities of sewerage as well as dangerous industrial effluents. Roads and rail tracks disappeared under waist-high and even chest-high floodwaters; in the northern part of the city, where the rainfall was largely concentrated, entire neighborhoods were inundated: 2.5 million people 'were under water for hours together.'

On weekdays Mumbai's suburban railway network transports close to 6.6 million passengers; buses carry more than 1.5 million. The deluge came down on a Tuesday, beginning at around 2 p.m. Local train services were soon disrupted, and by 4:30 p.m. none were moving; several arterial roads and intersections were cut off by floodwaters at about the same time. The situation worsened as more and more vehicles poured on to the roads; in many parts of the city traffic came to a complete standstill. Altogether two hundred kilometers of road were submerged; some motorists drowned in their cars because short-circuited electrical systems would not allow them to open doors and windows. Thousands of scooters, motorcycles, cars, and buses were abandoned on the water-logged roads.

At around 5 p.m. cellular networks failed; most landlines stopped working too. Soon much of the city's power supply was also cut off (although not before several people had been electrocuted): parts of the city would remain without power for several days. Two million people, including many school children, were stranded, with no means of reaching home; a hundred and fifty thousand commuters were jammed into the city's two major railway stations. Those without money were unable to withdraw cash because ATM services had been knocked out as well.

Road, rail, and air services would remain cut off for

two days. Over five hundred people died: many were washed away in the floods; some were killed in a landslide. Two thousand residential buildings were partially or completely destroyed; more than ninety thousand shops, schools, health care centers, and other buildings suffered damage.

While Mumbai's poor, especially the inhabitants of some of its informal settlements, were among the worst affected, the rich and famous were not spared either. The most powerful politician in the city had to be rescued from his home in a fishing boat; many Bollywood stars and industrialists were stranded or trapped by floodwaters.

Through all of this the people of Mumbai showed great generosity and resilience, sharing food and water and opening up their homes to strangers. Yet, as one observer notes, on July 26, 2005, it became 'clear to many million people in Mumbai that life may never be quite the same again. An exceptional rainstorm finally put to rest the long prevailing myth of Mumbai's indestructible resilience to all kinds of shocks, including that of the partition.'

In the aftermath of the deluge, many recommendations were made by civic bodies, NGOs, and even the courts. But ten years later, when another downpour occurred on June 10, 2015, it turned out that few of the recommended measures had been implemented: even though the volume of rainfall was only a third of that of the deluge of 2005, many parts of the city were again swamped by floodwaters.

What does Mumbai's experience of the downpour of

2005 tell us about what might, or might not, happen if a major storm happens to hit the city? The events will, of course, unfold very differently: to start with, a cyclone will arrive not with a few hours' notice, as was the case with the deluges, but after a warning period of several days. Storms are now so closely tracked, from the time they form onward, that there is usually an interval of a few days when emergency measures can be put in place.

Of these emergency measures, probably the most effective is evacuation. In historically cyclone-prone areas, like eastern India and Bangladesh, systems have been set up to move millions of people away from the coast when a major storm approaches; these measures have dramatically reduced casualties in recent years. But the uptick in cyclonic activity in the Arabian Sea is so recent that there has yet been no need for large-scale evacuations on the subcontinent's west coast. Whether such evacuations could be organized is an open question. Mumbai has been lucky not to have been hit by a major storm in more than a century; perhaps for that reason the possibility appears not to have been taken adequately into account in planning for disasters. Moreover, here, 'as in most megacities, disaster management is focused on post-disaster response.'

In Mumbai disaster planning seems to have been guided largely by concerns about events that occur with little or no warning, like earthquakes and deluges: evacuations usually follow rather than precede disasters of this kind. With a cyclone, given a lead-up period of several days, it would not be logistically impossible to evacuate large parts of the city before the storm's arrival:

its rail and port facilities would certainly be able to move millions of people to safe locations on the mainland. But in order to succeed, such an evacuation would require years of planning and preparation; people in at-risk areas would also need to be educated about the dangers to which they might be exposed. And that exactly is the rub – for in Mumbai, as in Miami and many other coastal cities, these are often the very areas in which expensive new construction projects are located. Property values would almost certainly decline if residents were to be warned of possible risks – which is why builders and developers are sure to resist efforts to disseminate disaster-related information. One consequence of the last two decades of globalization is that real estate interests have acquired enormous power, not just in Mumbai but around the world; very few civic bodies, especially in the developing world, can hope to prevail against construction lobbies, even where it concerns public safety. The reality is that 'growth' in many coastal cities around the world now depends on ensuring that a blind eye is turned toward risk.

Even with extensive planning and preparation the evacuation of a vast city is a formidable task, and not only for logistical reasons. The experience of New Orleans, in the days before Hurricane Katrina, or of New York before Sandy, or the city of Tacloban before Haiyan, tells us that despite the most dire warnings large numbers of people will stay behind; even mandatory evacuation orders will be disregarded by many. In the case of a megacity like Mumbai this means that hundreds of thousands, if not millions, will find themselves in harm's way when a cyclone makes landfall. Many will no doubt assume that having dealt with the floods of the recent past they will also be able to ride out a storm.

But the impact of a Category 4 or 5 cyclone will be very different from anything that Mumbai has experienced in living memory. During the deluges of 2005 and 2015 rain fell heavily on some parts of the city and lightly on others: the northern suburbs bore the brunt of the rainfall in both cases. The effects of the flooding were also most powerfully felt in low-lying areas and by the residents of ground-level houses and apartments; people living at higher elevations, and on the upper stories of tall buildings, were not as badly affected.

But the winds of a cyclone will spare neither low nor high; if anything, the blast will be felt most keenly by those at higher elevations. Many of Mumbai's tall buildings have large glass windows; few, if any, are reinforced. In a cyclone these exposed expanses of glass will have to withstand, not just hurricane-strength winds, but also flying debris. Many of the dwellings in Mumbai's informal settlements have roofs made of metal sheets and corrugated iron; cyclone-force winds will turn these, and the thousands of billboards that encrust the city, into deadly projectiles, hurling them with great force at the glass-wrapped towers that soar above the city.

Nor will a cyclone overlook those parts of the city that were spared the worst of the floods; to the contrary they will probably be hit first and hardest. The cyclones that have struck the west coast of India in the past have all traveled upward on a northeasterly tack, from the southern quadrant of the Arabian Sea. A cyclone moving in this direction would run straight into south Mumbai, where many essential civic and national institutions are located.

The southernmost tip of Mumbai consists of a tongue of low-lying land, much of it reclaimed; several important military and naval installations are located there, as is one of the country's most important scientific bodies – the Tata Institute of Fundamental Research. A storm surge of two or three meters would put much of this area under water; single-story buildings may be submerged almost to the roof. And an even higher surge is possible.

Not far from here lie the areas in which the city's most famous landmarks and institutions are located: most notably, the iconic Marine Drive, with its sea-facing hotels, famous for their sunset views, and its necklace-like row of art deco buildings. All of this sits on reclaimed land; at high tide waves often pour over the seawall. A storm surge would be barely impeded as it swept over and advanced eastward.

A distance of about four kilometers separates south Mumbai's two sea-facing shorelines. Situated on the east side are the city's port facilities, the legendary Taj Mahal Hotel, and the plaza of the Gateway of India, which is already increasingly prone to flooding. Beyond lies a much-used fishing port: any vessels that had not been moved to safe locations would be seized by the storm surge and swept toward the Gateway of India and the Taj Hotel.

At this point waves would be pouring into South Mumbai from both its sea-facing shorelines; it is not inconceivable that the two fronts of the storm surge would meet and merge. In that case the hills and promontories of south Mumbai would once again become islands, rising out of a wildly agitated expanse of water. Also visible above the waves would be the upper stories of many of the city's most important institutions: the Town Hall, the state legislature, the Chhatrapati Shivaji Railway Terminus, the towering headquarters of the Reserve Bank of India, and the skyscraper that houses India's largest and most important stock exchange.

Much of south Mumbai is low lying; even after the passing of the cyclone many neighborhoods would probably be waterlogged for several days; this will be true of other parts of the city as well. If the roads and rail lines are cut for any length of time, food and water shortages may develop, possibly leading to civil unrest. In Mumbai waterlogging often leads to the spread of illness and disease: the city's health infrastructure was intended to cater to a population of about half its present size; its municipal hospitals have only forty thousand beds. Since many hospitals will have been evacuated before the storm, it may be difficult for the sick and injured to get medical attention. If Mumbai's stock exchange and Reserve Bank are rendered inoperative, then India's financial and commercial systems may be paralyzed.

But there is another possibility, yet more frightening. Of the world's megacities, Mumbai is one of the few that has a nuclear facility within its urban limits: the Bhabha Atomic Research Centre at Trombay. To the north, at Tarapur, ninety-four kilometers from the city's periphery, lies another nuclear facility. Both these plants sit right upon the shoreline, as do many other nuclear installations around the world: these locations were chosen in order to give them easy access to water.

With climate change many nuclear plants around the world are now threatened by rising seas. An article in the *Bulletin of the Atomic Scientists* notes: 'During massive storms . . . there is a greatly increased chance of the loss of power at a nuclear power plant, which significantly contributes to safety risks.' Essential cooling systems could fail; safety systems could be damaged; contaminants could seep into the plant and radioactive water could leak out, as happened at the Fukushima Daiichi plant.

What threats might a major storm pose for nuclear plants like those in Mumbai's vicinity? I addressed this question to a nuclear safety expert, M. V. Ramana, of the Program on Science and Global Security at Princeton University. His answer was as follows: 'My biggest concerns have to do with the tanks in which liquid radioactive waste is stored. These tanks contain, in high concentrations, radioactive fission products and produce a lot of heat due to radioactive decay; explosive chemicals can also be produced in these tanks, in particular hydrogen gas. Typically waste storage facilities include several safety systems to prevent explosions. During major storms, however, some or all of these systems could be simultaneously disabled: cascading failures could make it difficult for workers to carry out any repairs – this is assuming that there will be any workers available and capable of undertaking repairs during a major storm. An explosion at such a tank, depending on the energy of the explosion and the exact weather conditions, could lead to the dispersal of radio-activity over hundreds of square kilometers; this in turn could require mass evacuations or the long-term cessation of agriculture in regions of high contamination.'

Fortunately, the chances of a cyclone hitting Mumbai are small in any given year. But there is no doubt whatsoever of the threats that will confront the city because of other climate change impacts: increased precipitation and rising sea levels. If there are substantial increases in rainfall over the next few decades, as climate models predict, then damaging floods will become more frequent. As for sea levels, if they rise by a meter or more by the end of the century, as some climate scientists fear they might, then some parts of south Mumbai will gradually become uninhabitable.

A similar fate awaits two other colonial cities, founded in the same century as Mumbai: Chennai (Madras), which also experienced a traumatic deluge in 2015; and Kolkata, to which I have close familial links.

Unlike Chennai and Mumbai, Kolkata is not situated beside the sea. However, much of its surface area is below sea level, and the city is subject to regular flooding: like everyone who has lived in Kolkata, I have vivid memories of epic floods. But long familiarity with flooding tends to have a lulling effect, which is why it came as a shock to me when I learned, from a World Bank report, that Kolkata is one of the global megacities that is most at risk from climate change; equally shocking was the discovery that my family's house, where my mother and sister live, is right next to one of the city's most threatened neighborhoods.

The report forced me to face a question that eventually confronts everybody who takes the trouble to inform themselves about climate change: what can I do to protect my family and loved ones now that I know what lies ahead? My mother is elderly and increasingly frail; there is no telling how she would fare if the house were to be cut off by a flood and medical attention were to become unavailable for any length of time.

After much thought I decided to talk to my mother about moving. I tried to introduce the subject tactfully, but it made little difference: she looked at me as though I had lost my mind. Nor could I blame her: it *did* seem like lunacy to talk about leaving a beloved family home, with all its memories and associations, simply because of a threat outlined in a World Bank report.

It was a fine day, cool and sunlit; I dropped the subject.

But the experience did make me recognize something that I would otherwise have been loathe to admit: contrary to what I might like to think, my life is not guided by reason; it is ruled, rather, by the inertia of habitual motion. This is indeed the condition of the vast majority of human beings, which is why very few of us will be able to adapt to global warming if it is left to us, as individuals, to make the necessary changes; those who will uproot themselves and make the right preparations are precisely those obsessed monomaniacs who appear to be on the borderline of lunacy.

If whole societies and polities are to adapt then the necessary decisions will need to be made collectively, within political institutions, as happens in wartime or national emergencies. After all, isn't that what politics, in its most fundamental form, is about? Collective survival and the preservation of the body politic?

Yet, to look around the world today is to recognize that with some notable exceptions, like Holland and China, there exist very few polities or public institutions that are capable of implementing, or even contemplating, a managed retreat from vulnerable locations. For most governments and politicians, as for most of us as individuals, to leave the places that are linked to our memories and attachments, to abandon the homes that have given our lives roots, stability, and meaning, is nothing short of unthinkable.

12.

It is surely no accident that colonial cities like Mumbai, New York, Boston, and Kolkata were all brought into being through early globalization. They were linked to each other not only through the circumstances of their founding but also through patterns of trade that expanded and accelerated Western economies. These cities were thus the drivers of the very processes that now threaten them with destruction. In that sense, their predicament is but an especially heightened instance of a plight that is now universal.

It isn't only in retrospect that the siting of some of these cities now appear as acts of utter recklessness: Bombay's first Parsi residents were reluctant to leave older, more sheltered ports like Surat and Navsari and had to be offered financial incentives to move to the newly founded city. Similarly, Qing dynasty officials were astonished to learn that the British intended to build a city on the island of Hong Kong: why would anyone want to create a settlement in a place that was so exposed to the vagaries of the earth?

But in time, sure enough, there was a collective setting aside of the knowledge that accrues over generations through dwelling in a landscape. People began to move closer and closer to the water.

How did this come about? The same question arises also in relation to the coast around Fukushima, where stone tablets had been placed along the shoreline in the Middle Ages to serve as tsunami warnings; future generations were explicitly told 'Do not build your homes below this point!'

The Japanese are certainly no more inattentive to the words of their ancestors than any other people: yet not only did they build *exactly* where they had been warned *not* to, they actually situated a nuclear plant there.

This too is an aspect of the uncanny in the history of our relations with our environments. It is not as if we had not been warned; it is not as if we were ignorant of the risks. An awareness of the precariousness of human existence is to be found in every culture: it is reflected in biblical and Quranic images of the Apocalypse, in the figuring of the Fimbulwinter in Norse mythology, in tales of *pralaya* in Sanskrit literature, and so on. It was the literary imagination, most of all, that was everywhere informed by this awareness.

Why then did these intuitions withdraw, not just from the minds of the founders of colonial cities, but

also from the forefront of the literary imagination? Even
in the West, the earth did not come to be regarded as
moderate and orderly until long after the advent of modernity: for poets and writers, it was not until the late
nineteenth century that Nature lost the power to evoke
that form of terror and awe that was associated with the
'sublime.' But the practical men who ran colonies and
founded cities had evidently acquired their indifference
to the destructive powers of the earth much earlier.

How did this come about? How did a state of consciousness come into being such that millions of people
would move to such dangerously exposed locations?

The chronology of the founding of these cities creates
an almost irresistible temptation to point to the European Enlightenment's predatory hubris in relation to the
earth and its resources. But this would tell us very little
about the thinking of the men who built and planned
that base in the Nicobars: if hubris and predation had
anything to do with their choice of site, it was at a great
remove. Between them and the cartographers and surveyors of an earlier era there was, I think, a much more
immediate link: a habit of mind that proceeded by creating discontinuities; that is to say, they were trained to
break problems into smaller and smaller puzzles until a
solution presented itself. This is a way of thinking that
deliberately excludes things and forces ('externalities')
that lie beyond the horizon of the matter at hand: it is a
perspective that renders the interconnectedness of Gaia
unthinkable.

The urban history of Bengal provides an interesting
illustration of what I am trying to get at. Colonial

Calcutta, which was for a long time the capital of the British Raj, was founded on the banks of the Hooghly River in the late seventeenth century. It had not been in existence for long before it came to be realized that the river was silting up. By the early nineteenth century, the East India Company had decided in principle that a new port would be built at a location closer to the Bay of Bengal. A site was chosen in the 1840s; it lay some thirty-five miles to the southeast of Calcutta, on the banks of a river called Matla (which means 'crazed' or 'intoxicated' in Bengali).

At that time, there lived in Calcutta an Englishman by the name of Henry Piddington. A shipping inspector by profession, he dabbled promiscuously in literature, philology, and the sciences until his true calling was revealed to him by a treatise: *An Attempt to Develop the Law of Storms*, written by an American meteorologist, Col. Henry Reid. Published in 1838, the book was an ambitious study of the circular motion of tropical storms. Colonel Reid's book inspired a great passion in Piddington, and he devoted himself to the field for the rest of his life. It was he who coined the word *cyclone*, and it is for this that he is best remembered today. But Piddington's particular interest was the phenomenon of the storm surge (or 'storm wave' as it was then called): he would eventually compile a detailed account of storm surges along the coast of Bengal and the devastation they had caused.

Because of his familiarity with this subject, Piddington understood that the proposed port on the Matla River would be exposed to extreme cyclonic hazard.

Such was his alarm that in 1853 he published a pamphlet, addressed to the then governor-general, in which he issued this ominous warning: 'every one and everything must be prepared to see a day when, in the midst of the horrors of a hurricane, they will find a terrific mass of salt water rolling in, or rising up upon them, with such rapidity that the whole settlement will be inundated to a depth from five to fifteen feet.'

Piddington's warnings fell on deaf ears: to the builders and civil servants who were working on the new city, he must have sounded like a madman – in the measureable, discrete universes that they worked within there was no place for a phenomenon that was born hundreds of miles away and came storming over the seas like a 'wonderful meteor' (to use Piddington's words).

It was probably the very scale of the phenomenon invoked by Piddington that made it unthinkable to those eminently practical men, accustomed as they were to the 'regularity of bourgeois life.' The port continued to rise, even through the great uprising of 1857: it was built on a lavish scale, with banks, hotels, a railway station, and imposing public buildings. The city was formally inaugurated in 1864 with a grand ceremony: it was named Port Canning, after a former governor-general.

Port Canning's claims to grandeur were short-lived. A mere three years after its inauguration, it was struck by a cyclone, just as Piddington had predicted. And even though the accompanying storm surge was a modest one, rising to only six feet, it caused terrible destruction. The city was abandoned four years later

(Canning is now a small river port and access point for the Sundarbans). Piddington thus became one of the first Cassandras of climate science.

13.

If I have dwelt on this at some length, it is because the discontinuities that I have pointed to here have a bearing also on the ways in which worlds are created within novels. A 'setting' is what allows most stories to unfold; its relation to the action is as close as that of a stage to a play. When we read *Middlemarch* or *Buddenbrooks* or *Waterland*, or the great Bengali novel *A River Called Titash*, we enter into their settings until they begin to seem real to us; we ourselves become emplaced within them. This exactly is why 'a sense of place' is famously one of the great conjurations of the novel as a form.

What the settings of fiction have in common with sites measured by surveyors is that they too are constructed out of discontinuities. Since each setting is particular to itself, its connections to the world beyond are inevitably made to recede (as, for example, with the imperial networks that make possible the worlds portrayed by Jane Austen and Charlotte Brontë). Unlike epics, novels do not usually bring multiple universes into conjunction; nor are their settings transportable outside their context in the manner of, say, the Ithaca of the *Odyssey* or the Ayodhya of the *Ramayana*.

In fiction, the immediate discontinuities of place are nested within others: Maycomb, Alabama, the setting of

To Kill a Mockingbird, becomes a stand-in for the whole of the Deep South; the *Pequod*, the small Nantucket whaling vessel in *Moby Dick*, becomes a metaphor for America. In this way, settings become the vessel for the exploration of that ultimate instance of discontinuity: the nation-state.

In novels discontinuities of space are accompanied also by discontinuities of time: a setting usually requires a 'period'; it is actualized within a certain time horizon. Unlike epics, which often range over eons and epochs, novels rarely extend beyond a few generations. The *longue durée* is not the territory of the novel.

It is through the imposition of these boundaries, in time and space, that the world of a novel is created: like the margins of a page, these borders render places into texts, so that they can be read. The process is beautifully illustrated in the opening pages of *A River Called Titash*. Published in 1956, this remarkable novel was the only work of fiction published by Adwaita Mallabarman, who belonged to an impoverished caste of Dalit fisherfolk. The novel is set in rural Bengal, in a village on the shores of a fictional river, Titash.

Bengal is, as I have said, a land of titanic rivers. Mallabarman gestures toward the vastness of the landscape with these words: 'The bosom of Bengal is draped with rivers and their tributaries, twisted and intertwined like tangled locks, streaked with the white of foamy waves.'

But almost at once he begins to detach the setting of his novel from the larger landscape: all rivers are not the same, he tells us, some are like 'a frenzied sculptor at work, destroying and creating restlessly in crazed joy, riding the high-flying swing of fearsome energy – here is one kind of art.'

And then, in a striking passage, the writer announces his own intentions and premises: 'There is another kind of art ... The practitioner of this art cannot depict Mahakaal (Shiva the Destroyer) in his cosmic dance of creation and destruction – the awesome vision of tangled brown hair tumbling out of the coiled mass will not come from this artist's brush. The artist has come away from the rivers Padma, Meghna, and Dhaleswari to find a home beside Titash.

'The pictures this artist draws please the heart. Little villages dot the edges of the water. Behind these villages are stretches of farmland.'

This is how the reader learns that the Titash, although it is a part of a landscape of immense waterways, is itself a small, relatively gentle river: 'No cities or large towns ever grew up on its banks. Merchant boats with giant sails do not travel its waters. Its name is not in the pages of geography books.'

In this way, through a series of successive exclusions, Mallabarman creates a space that will submit to the techniques of a modern novel: the rest of the landscape is pushed farther and farther into the background until at last we have a setting that can carry a narrative. The setting becomes, in a sense, a self-contained ecosystem, with the river as the sustainer both of life and of the narrative. The impetus of the novel, and its poignancy, come from the Titash itself: it is the river's slow drying up that directs the lives of the characters. The Titash is, of course, but one strand of the 'tangled locks' of an immense network of rivers and its flow is necessarily ruled by the dynamics of the landscape. But it is precisely by excluding those inconceivably large forces, and by telescoping the changes into the duration of a limited-time horizon, that the novel becomes narratable.

Contrast this with the universes of boundless time and space that are conjured up by other forms of prose narrative. Here, for example, are a couple of passages from the beginning of the sixteenth-century Chinese folk epic *The Journey to the West*: 'At this point the firmament first acquired its foundation. With another 5,400 years came the Tzu epoch; the ethereal and the light rose up to form the four phenomena of the sun, the moon, the stars, and the heavenly bodies . . . Following P'an Ku's construction of the universe . . . the world was divided into four great continents . . . Beyond the ocean there was a country named Ao-lai. It was near a great ocean, in the midst of which was located the famous Flower-Fruit Mountain.'

Here is a form of prose narrative, still immensely popular, that ranges widely and freely over vast expanses of time and space. It embraces the inconceivably large almost to the same degree that the novel shuns it. Novels, on the other hand, conjure up worlds that become real precisely because of their finitude and distinctiveness. Within the mansion of serious fiction, no one will speak of how the continents were created; nor will they refer to the passage of thousands of years: connections and events on this scale appear not just unlikely but also absurd within the delimited horizon of a novel – when they intrude, the temptation to lapse into satire, as in Ian McEwan's *Solar*, becomes almost irresistible.

But the earth of the Anthropocene is precisely a world of insistent, inescapable continuities, animated by forces that are nothing if not inconceivably vast. The waters that are invading the Sundarbans are also swamping Miami Beach; deserts are advancing in China as well as Peru; wildfires are intensifying in Australia as well as Texas and Canada.

There was never a time, of course, when the forces of weather and geology did *not* have a bearing on our lives – but neither has there ever been a time when they have pressed themselves on us with such relentless directness. We have entered, as Timothy Morton says, the age of hyperobjects, which are defined in part by their stickiness, their ever-firmer adherence to our lives: even to speak of the weather, that safest of subjects, is now to risk a quarrel with a denialist neighbor. No less than they mock the discontinuities and boundaries of the nation-state do these connections defy the boundedness of 'place,' creating continuities of experience between Bengal and Louisiana, New York and Mumbai, Tibet and Alaska.

I was recently sent a piece about a mangrove forest in Papua New Guinea. This was once a 'place' in the deepest sense that it was linked to its inhabitants through a dense web of mutual sustenance and symbolism. But in the wet season of 2007, 'the barrier beaches were breached, cutting innumerable channels through to the lakes. Sand poured through them. Tidal surges tore across the villages, leaving behind a spectacle of severed trunks of coconut palms and dead shoreline trees, drifting canoes, trenches, and gullies. Entire villages had to be evacuated.' Eventually the inhabitants were forced to abandon their villages.

The Anthropocene has reversed the temporal order of modernity: those at the margins are now the first to experience the future that awaits all of us; it is they who confront most directly what Thoreau called 'vast, Titanic, inhuman nature.' Nor is it any longer possible to exclude this dynamic even from places that were once renowned for their distinctiveness. Can anyone write about Venice any more without mentioning the *aqua alta*, when the waters of the lagoon swamp the city's streets and courtyards? Nor can they ignore the relationship that this has with the fact that one of the languages most frequently heard in Venice is Bengali: the men who run the quaint little vegetable stalls and bake the pizzas and even play the accordion are largely Bangladeshi, many of them displaced by the same phenomenon that now threatens their adopted city – sea-level rise.

Behind all of this lie those continuities and those inconceivably vast forces that have now become impossible to exclude, even from texts.

Here, then, is another form of resistance, a scalar one, that the Anthropocene presents to the techniques that are most closely identified with the novel: its essence consists of phenomena that were long ago expelled from the territory of the novel – forces of unthinkable magnitude that create unbearably intimate connections over vast gaps in time and space.

14.

I would like to return, for a moment, to the images I started with: of apparently inanimate things coming suddenly alive. This, as I said earlier, is one of the uncanniest effects of the Anthropocene, this renewed awareness of the elements of agency and consciousness that humans share with many other beings, and even perhaps the planet itself.

But such truth as this statement has is only partial: for the fact is that a great number of human beings had never lost this awareness in the first place. In the Sundarbans, for example, the people who live in and around the mangrove forest have never doubted that tigers and many other animals possess intelligence and agency. For the first peoples of the Yukon, even glaciers are endowed with moods and feelings, likes and dislikes. Nor would these conceptions have been unthinkable for a scientist like Sir Jagadish Chandra Bose (1858–1937), who attributed elements of consciousness to vegetables and even metals, or for the primatologist Imanishi Kinji (1902–92) who insisted on 'the unity of all elements on the planet earth – living and non-living.'

Neither is it the case that we were all equally captive to Cartesian dualism before the awareness of climate change dawned on us: my ancestors were certainly not in its thrall, and even I was never fully acculturated to that view of the world. Indeed, I would venture to say that this is true for most people in the world, even in the West. To the great majority of people everywhere, it has always been perfectly evident that dogs, horses, elephants, chimpanzees, and many other animals possess intelligence and emotions. Did anyone ever really believe, *pace* Descartes, that animals are automatons? 'Surely Descartes never saw an ape' wrote Linnaeus, who found it no easy matter to draw a line between human and animal. Even the most devoted Cartesian will probably have no difficulty in interpreting the emotions of a dog that has backed him up against a wall.

Nowhere is the awareness of nonhuman agency more evident than in traditions of narrative. In the Indian epics – and this is a tradition that remains vibrantly alive to this day – there is a completely matter-of-fact acceptance of the agency of nonhuman beings of many kinds. I refer not only to systems of belief but also to techniques of storytelling: nonhumans provide much of the momentum of the epics; they create the resolutions that allow the narrative to move forward. In the *Iliad* and the *Odyssey* too the intervention of gods, animals, and the elements is essential to the machinery of narration. This is true for many other narrative traditions as well, Asian, African, Mediterranean, and so on. The Hebrew Bible is no exception; as the theologian Michael Northcott points out, 'At the heart of Judaism is a God who is encountered through Nature and events rather than words or texts. Christianity, by contrast, and then Islam, is a form of religion that is less implicated in the weather, climate and political power and more invested in words and texts.'

But even within Christianity, it was not till the advent of Protestantism perhaps that Man began to dream of achieving his own self-deification by radically isolating himself before an arbitrary God. Yet that dream of silencing the nonhuman has never been completely realized, not even within the very heart of contemporary modernity; indeed, it would seem that one aspect of the agency of nonhumans is their uncanny ability to stay abreast of technology. Even among today's teenagers and twenty-somethings, whose most intimate familiars are man-made objects like iPads and iPhones, an awareness evidently still lingers that elements of agency are concealed everywhere within our surroundings: why else should the charts of best-selling books and top-grossing films continue to be so heavily weighted in favor of those that feature werewolves, vampires, witches, shape-shifters, extraterrestrials, mutants, and zombies?

So the real mystery in relation to the agency of nonhumans lies not in the renewed recognition of it, but rather in how this awareness came to be suppressed in the first place, at least within the modes of thought and expression that have become dominant over the last couple of centuries. Literary forms have clearly played an important, perhaps critical, part in the process. So, if for a moment we were to take seriously the premise that I started with – that the Anthropocene has forced us to recognize that there are other, fully aware eyes looking over our shoulders – then the first question to present itself is this: What is the place of the nonhuman in the modern novel?

To attempt an answer is to confront another of the uncanny effects of the Anthropocene: it was in exactly the period in which human activity was changing the earth's atmosphere that the literary imagination became radically centered on the human. Inasmuch as the non-human was written about at all, it was not within the mansion of serious fiction but rather in the outhouses to which science fiction and fantasy had been banished.

15.

The separation of science fiction from the literary mainstream came about not as the result of a sudden drawing of boundaries but rather through a slow and gradual process. There was, however, one moment that was critical to this process, and it happens to have had a link to a climate-related event.

The seismic event that began on April 5, 1815, on Mount Tambora, three hundred kilometers to the east of Bali, was the greatest volcanic eruption in recorded history. Over the next few weeks, the volcano would send one hundred cubic kilometers of debris shooting into the air. The plume of dust – 1.7 million tons of it – soon spread around the globe, obscuring the sun and causing temperatures to plunge by three to six degrees. There followed several years of severe climate disruption; crops failed around the world, and there were famines in Europe and China; the change in temperature may also have triggered a cholera epidemic in India. In many parts of the world, 1816 would come to be known as the 'Year without a Summer.'

In May that year, Lord Byron, besieged by scandal, left England and moved to Geneva. He was accompanied by his physician, John Polidori. As it happened, Percy Bysshe Shelley and Mary Wollstonecraft Godwin, who had recently eloped together, were also in Geneva at the time, staying at the same hotel. Accompanying them was Mary Godwin's stepsister, Claire, with whom Byron had had a brief affair in England.

Shelley and Byron met on the afternoon of May 27, and shortly afterward they moved, with their respective parties, to two villas on the shores of Lake Geneva. From there they were able to watch thunderstorms approaching over the mountains. 'An almost perpetual rain confines us principally to the house,' Mary Shelley wrote. 'One night we *enjoyed* a finer storm than I had ever before beheld. The lake was lit up, the pines on the Jura made visible, and all the scene illuminated for an instant, when a pitchy blackness succeeded, and the thunder came in frightful bursts over our heads amid the darkness.'

One day, trapped indoors by incessant rain, Byron suggested that they all write ghost stories. A few days later, he outlined an idea for a story 'on the subject of the vampyric aristocrat, August Darvell.' After eight pages, Byron abandoned the story, and his idea was taken up instead by Polidori: it was eventually published as *The Vampyre* and is now regarded as the first in an ever-fecund stream of fantasy writing.

Mary Shelley too had decided to write a story, and one evening (a stormy one no doubt), the conversation turned to the question of whether 'a corpse would be reanimated: galvanism had given token of such things: perhaps the component parts of a creature might be manufactured, brought together and endowed with vital warmth.' The next day, she began writing *Franken-stein, or The Modern Prometheus*. Published in 1818, the novel created a sensation: it was reviewed in the best-known journals, by some of the most prominent writers of the time. Sir Walter Scott wrote an enthusiastic review, and he would say later that he preferred it to his own novels. At that time, there does not seem to have been any sense that *Frankenstein* belonged outside the literary mainstream; only later would it come to be regarded as the first great novel of science fiction.

Although Byron never did write a ghost story, he did compose a poem called 'Darkness,' which was imbued with what we might today call 'climate despair':

> The world was void,
> The populous and the powerful – was a lump,
> Seasonless, herbless, treeless, manless, lifeless –
> A lump of death – a chaos of hard clay.
> The rivers, lakes, and ocean all stood still,
> And nothing stirred within their silent depths.

Reflecting on the 'wet, ungenial summer' of 1816 and its role in the engendering of these works, Geoffrey Parker writes, 'All three works reflect the disorientation and desperation that even a few weeks of abrupt climate change can cause. Since the question today is not *whether* climate change will strike some part of our planet again, but *when*, we might re-read Byron's poem as we choose.'

16.

To ask how science fiction came to be demarcated from the literary mainstream is to summon another question: What is it in the nature of modernity that has led to this separation? A possible answer is suggested by Bruno Latour, who argues that one of the originary impulses of modernity is the project of 'partitioning,' or deepening the imaginary gulf between Nature and Culture: the former comes to be relegated exclusively to the sciences and is regarded as being off-limits to the latter.

Yet, to look back at the evolution of literary culture from this vantage point is to recognize that the project of partitioning has always been contested, and never more so than at the inception, and nowhere more vigorously than in places that were in the vanguard of modernity. As proof of this, we have only to think of William Blake, asking of England:

> And was Jerusalem builded here,
> Among these dark Satanic mills?

And of Wordsworth's sonnet, 'The World Is Too Much With Us':

> Little we see in Nature that is ours;
> We have given our hearts away, a sordid boon! . . .
> . . . Great God! I'd rather be

A Pagan suckled in a creed outworn;
So might I, standing on this pleasant lea,
Have glimpses that would make me less forlorn.

Nor was it only in England, but also throughout Europe and North America that partitioning was resisted, under the banners variously of romanticism, pastoralism, transcendentalism, and so on. Poets were always in the forefront of the resistance, in a line that extends from Hölderlin and Rilke to such present-day figures as Gary Snyder and W. S. Merwin.

But being myself a writer of fiction, it is the novel that interests me most, and when we look at the evolution of the form, it becomes evident that its absorption into the project of partitioning was presaged already in the line of Wordsworth's that I quoted above: 'I'd rather be / A Pagan suckled in a creed outworn.'

It is with these words that the poet, even as he laments the onrushing intrusion of the age, announces his surrender to the most powerful of its tropes: that which envisages time as an irresistible, irreversible forward movement. This jealous deity, the Time-god of modernity, has the power to decide who will be cast into the shadows of backwardness – the dark tunnel of time 'outworn' – and who will be granted the benediction of being ahead of the rest, always *en avant*. It is this conception of time (which has much in common with both Protestant and secular teleologies, like those of Hegel and Marx) that allows the work of partitioning to proceed within the novel, always aligning itself with the avant-garde as it hurtles forward in its impatience to erase every archaic reminder of Man's kinship with the nonhuman.

The history of this partitioning is, of course, an epic in itself, offering subplots and characters to suit the tastes of every reader. Here I want to dwell, for a moment, on a plot that completely reverses itself between the eighteenth century and today: the story of the literary tradition's curious relationship with science.

At the birth of modernity, the relationship between literature and science was very close and was perhaps perfectly exemplified in the figure of the writer Bernardin de Saint-Pierre, who wrote one of the earliest of best sellers, *Paul et Virginie*. Saint-Pierre regarded himself as primarily a naturalist and saw no conflict between his calling as writer and man of science. It is said of him that when taken to see the cathedral of Chartres, as a boy, he noticed nothing but the jackdaws that were roosting on the towers.

Goethe also famously saw no conflict between his literary and scientific interests, conducting experiments in optics, and propounding theories that remain compelling to this day. Herman Melville too was deeply interested in the study of marine animals and his views on the subject are, of course, expounded at length in *Moby Dick*. I could cite many other instances ranging from the mathematics of *War and Peace* to the chemistry of *Alice in Wonderland*, but there is no need: it is hardly a matter of dispute that Western writers remained deeply engaged with science through the nineteenth century.

Nor was this a one-sided engagement. Naturalists and scientists not only read but also produced some of the most significant literary works of the nineteenth century, such as Darwin's *Voyage of the Beagle* and Alfred Russell Wallace's *The Malay Archipelago*. Their works, in turn, served as an inspiration to a great number of poets and writers, including Tennyson.

How, then, did the provinces of the imaginative and the scientific come to be so sharply divided from each other? According to Latour the project of partitioning is supported always by a related enterprise: one that he describes as 'purification,' the purpose of which is to ensure that Nature remains off-limits to Culture, the knowledge of which is consigned entirely to the sciences. This entails the marking off and suppression of hybrids – and that, of course, is exactly the story of the branding of science fiction, as a genre *separate* from the literary mainstream. The line that has been drawn between them exists only for the sake of neatness; because the zeitgeist of late modernity could not tolerate Nature–Culture hybrids.

Nor is this pattern likely to change soon. I think it can be safely predicted that as the waters rise around us, the mansion of serious fiction, like the doomed waterfront properties of Mumbai and Miami Beach, will double down on its current sense of itself, building ever higher barricades to keep the waves at bay.

The expulsion of hybrids from the manor house has long troubled many who were thus relegated to the status of genre writers, and rightly so, for nothing could be more puzzling than the strange conceit that science fiction deals with material that is somehow contaminated; nothing could better express the completeness of the literary mainstream's capitulation to the project of partitioning. And this capitulation has come at a price, for it is literary fiction itself that has been diminished by it. If a list were to be made of the late twentieth-century novelists whose works remain influential today, we would find, I suspect, that many who once bestrode the literary world like colossi are entirely forgotten while writers like Arthur C. Clarke, Raymond Bradbury, and Philip K. Dick are near the top of the list.

That said, the question remains: Is it the case that science fiction is better equipped to address the Anthropocene than mainstream literary fiction? This might appear obvious to many. After all, there is now a new genre of science fiction called 'climate fiction' or cli-fi. But cli-fi is made up mostly of disaster stories set in the future, and that, to me, is exactly the rub. The future is but one aspect of the Anthropocene: this era also includes the recent past, and, most significantly, the present.

In a perceptive essay on science fiction and speculative fiction, Margaret Atwood writes of these genres that they 'all draw from the same deep well: those imagined other worlds located somewhere apart from our everyday one: in another time, in another dimension, through a doorway into the spirit world, or on the other side of the threshold that divides the known from the unknown. Science Fiction, Speculative Fiction, Sword and Sorcery Fantasy, and Slipstream Fiction: all of them might be placed under the same large "wonder tale" umbrella.'

This lays out with marvelous clarity some of the ways in which the Anthropocene resists science fiction: it is precisely not an imagined 'other' world apart from ours;

nor is it located in another 'time' or another 'dimension.' By no means are the events of the era of global warming akin to the stuff of wonder tales; yet it is also true that in relation to what we think of as normal now, they are in many ways uncanny; and they have indeed opened a doorway into what we might call a 'spirit world' – a universe animated by nonhuman voices.

If I have been at pains to speak of resistances rather than insuperable obstacles, it is because these challenges can be, and have been, overcome in many novels: Liz Jensen's *Rapture* is a fine example of one such; another is Barbara Kingsolver's wonderful novel *Flight Behavior*. Both are set in a time that is recognizable as our own, and they both communicate, with marvelous vividness, the uncanniness and improbability, the magnitude and interconnectedness of the transformations that are now under way.

17.

Global warming's resistance to the arts begins deep underground, in the recesses where organic matter undergoes the transformations that make it possible for us to devour the sun's energy in fossilized forms. Think of the vocabulary that is associated with these substances: *naphtha, bitumen, petroleum, tar,* and *fossil fuels*. No poet or singer could make these syllables fall lightly on the ear. And think of the substances themselves: coal and the sooty residue it leaves on everything it touches; and petroleum – viscous, foul smelling, repellant to all the senses.

Of coal at least it can be said that the manner of its extraction is capable of sustaining stories of class solidarity, courage, and resistance, as in Zola's *Germinal*, for instance, and John Sayles's fine film *Matewan*.

The very materiality of coal is such as to enable and promote resistance to established orders. The processes through which it is mined and transported to the surface create an unusual degree of autonomy for miners; as Timothy Mitchell observes, 'the militancy that formed in these workplaces was typically an effort to defend this autonomy.' It is no coincidence, then, that coal miners were in the front lines of struggles for the expansion of political rights from the late nineteenth until the mid-twentieth century, and even afterward. It could even be argued that miners, and therefore the economy of coal itself, were largely responsible for the unprecedented expansion of democratic rights that occurred in the West between 1870 and the First World War.

The materiality of oil is very different from that of coal: its extraction does not require large numbers of workers, and since it can be piped over great distances, it does not need a vast workforce for its transportation and distribution. This is probably why its effects, politically speaking, have been the opposite of those of coal. That this might be the case was well understood by Winston Churchill and other leaders of the British and American political elites, which was why they went to great lengths to promote the large-scale use of oil. This effort gained in urgency after the historic strikes of the 1910s and '20s, in which miners, and workers who transported and distributed coal, played a major role; indeed, fear of working-class militancy was one of the reasons why a large part of the Marshall Plan's funds went toward effecting the switch from coal to oil. 'The corporatised democracy of postwar Western Europe was to be built,' as Mitchell notes, 'on this reorganisation of energy flows.'

For the arts, oil is inscrutable in a way that coal never was: the energy that petrol generates is easy to aestheticize – as in images and narratives of roads and cars – but the substance itself is not. Its sources are mainly hidden from sight, veiled by technology, and its workers are hard to mythologize, being largely invisible. As for the places where oil is extracted, they possess nothing of the raw visual power that is manifest, for example, in the mining photographs of Sebastião Salgado. Oil refineries are usually so heavily fortified that little can be seen of them other than a distant gleam of metal, with tanks, pipelines, derricks, glowing under jets of flame.

One such fortress figures in my first novel, *The Circle of Reason* (1986), a part of which is about the discovery of oil in a fictional emirate called al-Ghazira: 'out of the sand, there suddenly arose the barbed-wire fence of the Oiltown. From the other side of the fence, faces stared silently out – Filipino faces, Indian faces, Egyptian faces, Pakistani faces, even a few Ghaziri faces, a whole world of faces.'

Behind these eerie, dislocated enclaves of fenced-in faces and towering derricks lies a history that impinges on every life on this planet. This is true especially in regard to the Arabian peninsula, where oil brought about an encounter with the West that has had consequences that touch upon every aspect of our existence, extending from matters of security to the buildings that surround us and the quality of the air we breathe. Yet the strange reality is that this historic encounter, whose tremors and aftershocks we feel every day, has almost no presence in our imaginative lives, in art, music, dance, or literature.

Long after the publication of *The Circle of Reason*, I wrote a piece in which I attempted to account for this mysterious absence: 'To the principal protagonists in the Oil Encounter (which means, in effect, America and Americans, on the one hand, and the peoples of the Arabian peninsula and the Persian Gulf, on the other), the history of oil is a matter of embarrassment verging on the unspeakable, the pornographic. It is perhaps the one cultural issue on which the two sides are in complete agreement ... Try and imagine a major American writer taking on the Oil Encounter. The idea is literally inconceivable.'

The above passage figures in a review of one of the few works of fiction to address the Oil Encounter, a five-part series of novels by the Jordanian-born writer Abdel Rahman Munif. My review, entitled 'Petrofiction,' dealt only with the first two books in the cycle, which were published in English translation as *Cities of Salt* (*Mudun al Malh*) and *The Trench* (*Al-ukhdud*).

'The truth is,' I wrote in my review, 'that we do not yet possess the form that can give the Oil Encounter a literary expression. For this reason alone *Cities of Salt* . . . ought to be regarded as a work of immense significance.

It so happens that the first novel in the cycle is also in many ways a wonderful work of fiction, perhaps even in parts a great one.'

This was written in 1992. I was not aware then that *Cities of Salt* had been reviewed four years earlier by one of the most influential figures in the American literary firmament: John Updike. His review, when I read it, made a great impression on me: I found that in the process of writing about Munif's book Updike had also articulated, elegantly and authoritatively, a conception of the novel that was indisputably an accurate summing-up of a great deal of contemporary fiction. Yet it was a conception with which I found myself completely at odds.

The differences between Updike's views and mine have an important bearing on some of the aspects of the Anthropocene that I have been addressing here, so it is best to let him speak for himself. Here is what he had to say about *Cities of Salt:* 'It is unfortunate, given the epic potential of his topic, that Mr. Munif . . . appears to be . . . insufficiently Westernized to produce a narrative that feels much like what we call a novel. His voice is that of a campfire explainer; his characters are rarely fixed in our minds by a face or a manner or a developed motivation; no central figure develops enough reality to attract our sympathetic interest; and, this being the first third of a trilogy, what intelligible conflicts and possibilities do emerge remain serenely unresolved. There is almost none of that sense of individual moral adventure – of the evolving individual in varied and roughly equal battle with a world of circumstance – which since *Don Quixote* and *Robinson Crusoe,* has distinguished the novel from the fable and the chronicle; *Cities of Salt* is concerned, instead, with men in the aggregate.'

This passage is remarkable, in the first instance, because the conception of the novel that is articulated here is rarely put into words, even though it has come to exercise great sway across much of the world and especially in the Anglosphere. My own disagreement with it hinges upon the phrase that Updike uses to distinguish the novel from the fable and the chronicle: 'individual moral adventure.'

But why, I find myself asking, should the defining adventures of the novel be described as 'moral,' as opposed to, say, intellectual or political or spiritual? In what sense could it be said that *War and Peace* is about individual moral adventures? It is certainly true that some threads in the narrative could be described in this way, but they would account for only a small part of the whole. As for Tolstoy's own vision of what he had set out to do, he was emphatic that *War and Peace* was 'not a novel, still less a long poem, and even less a historical chronicle': his intention was to supersede and incorporate preceding forms – an ambition that can be seen also in Melville's *Moby Dick*. To fit these works within the frame of 'individual moral adventures' is surely a diminishment of the writers' intentions.

Clearly, it is in the word *moral* that the conundrum lies: What exactly does it mean? Is it intended perhaps to incorporate the senses also of the 'political,' the 'spiritual,' and the 'philosophical'? And if so, would not a question arise as to whether a single word can bear so great a burden?

I ask these questions not in order to parse small semantic differences. I believe that Updike had actually put his finger on a very important aspect of contemporary culture. I will return to this later, but for now I'd like to turn to another aspect of Updike's mapping of the territory of the novel: that which is excluded from it.

Updike draws this boundary line with great clarity: the reason why *Cities of Salt* does not feel 'much like a novel,' he tells us, is that it is concerned not with a sense of individual moral adventures but rather with 'men in the aggregate.' In other words, what is banished from the territory of the novel is precisely the collective.

But is it actually the case that novelists have shunned men (or women) in the aggregate? And inasmuch as they have, is it a matter of intention or narrative expediency? Charlotte Brontë's view, expressed in a letter to a critic, is worth noting: 'is not the real experience of each individual very limited?' she asks, 'and if a writer dwells upon that solely or principally is he not in danger of being an egotist?'

In a perceptive discussion of Updike's review, the critic Rob Nixon points out that Munif is 'scarcely alone in working with a crowded canvas and with themes of collective transformation'; Émile Zola, Upton Sinclair, and many others have also treated 'individual character as secondary to collective metamorphosis.'

Indeed, so numerous are the traces of the collective within the novelistic tradition that anyone who chose to look for them would soon be overwhelmed. Such being the case, should Updike's view be summarily dismissed? My answer is: no – because Updike was, in a certain sense, right. It is a fact that the contemporary novel has become ever more radically centered on the individual psyche while the collective – 'men in the aggregate' – has receded, both in the cultural and the fictional imagination. Where I differ from Updike is that I do not think that this turn in contemporary fiction has anything to do with the novel as a form: it is a matter of record that historically many novelists from Tolstoy and Dickens to Steinbeck and Chinua Achebe have written very effectively about 'men in the aggregate.' In many parts of the world, they continue to do so even now.

What Updike captures, then, is not by any means an essential element of the novel as a form; his characterization is true rather of a turn that fiction took at a certain time in the countries that were then leading the way to the 'Great Acceleration' of the late twentieth century. It is certainly no coincidence that these were the very places where, as Guy Debord observed, the reigning economic system was not only founded on isolation, it was also 'designed to produce isolation.'

I say it is no coincidence for two reasons. The first is that the acceleration in carbon emissions and the turn away from the collective are both, in one sense, effects of that aspect of modernity that sees time (in Bruno Latour's words) as 'an irreversible arrow, as capitalization, as progress.' I've noted before that this idea of a continuous and irreversible forward movement, led by an avant-garde, has been one of the animating forces of the literary and artistic imagination since the start of the twentieth century. A progression of this sort inevitably creates winners and losers, and in the case of twentieth-century fiction, one of the losers was exactly writing of the kind in which the collective had a powerful presence. Fiction of this sort was usually of a realist variety, and it receded because it was consigned to the netherworld of 'backwardness.'

But the era of global warming has made audible a new, nonhuman critical voice that forces us to ask whether those old realists were so 'used-up' after all. Consider the example of John Steinbeck, never a favorite of the avant-garde, and once famously dismissed by Lionel Trilling as a writer who thought 'like a social function, not a novelist.' Yet, if we look back upon Steinbeck now, in full awareness of what is now known about the future of the planet, his work seems far from superseded; quite the contrary. What we see, rather, is a visionary placement of the human within the nonhuman; we see a form, an approach that grapples with climate change avant la lettre.

Around the world too there are many writers – not all of them realists – from whose work neither the aggregate nor the nonhuman have ever been absent. To cite only a few examples from India: in Bengali, there is the work of Adwaita Mallabarman and Mahasweta Devi; in Kannada, Sivarama Karanth; in Oriya, Gopinath Mohanty; in Marathi, Vishwas Patil. Of these writers too I suspect that Updike would have said that their books were not much 'like what we call a novel.'

But once again, the last laugh goes to that sly critic, the Anthropocene, which has muddied, and perhaps even reversed, our understanding of what it means to be 'advanced.' Were we to adopt the arrow-like time perspective of the moderns, there is a sense in which we might even say of writers like Munif and Karanth that they were actually 'ahead' of their peers elsewhere.

Here, then, is another reason why something more than mere chance appears to be at play in the turn that fiction took as emissions were rising in the late twentieth century. It is one of the many turns of that period that give, in retrospect, the uncanny impression that global warming has long been toying with humanity (thus, for example, the three postwar decades, when emissions grew sharply, saw a *stabilization* of global temperatures). Similarly, at exactly the time when it has become clear that global warming is in every sense a collective predicament, humanity finds itself in the thrall of a dominant culture in which the idea of the collective has been exiled from politics, economics, and literature alike.

Inasmuch as contemporary fiction is caught in this thralldom, this is one of the most powerful ways in which global warming resists it: it is as if the gas had run out on a generation accustomed to jet skis, leaving them with the task of reinventing sails and oars.

18.

Mrauk-U is the site of a vast and enchanting complex of Buddhist pagodas and monasteries in western Burma. Once the capital of the Arakan (Rakhine) kingdom, it flourished between the fourteenth and seventeenth centuries. During that time, it was an important link in the networks of trade that spanned the Indian Ocean. Chinese ceramics and Indian textiles passed through it in quantity; merchants and travelers came from Gujarat, Bengal, East Africa, Yemen, Portugal, and China to sojourn in the city. The wealth they brought in allowed the kingdom's rulers to honor their religion by embarking on vast building projects. The site they created is smaller than Bagan or Angkor Wat, but it is, in its own way, just as interesting.

Getting to Mrauk-U isn't easy. The nearest town of any size is Sittwe (formerly Akyab), and from there the journey to the site can take a day or more, depending on the condition of the road. As Mrauk-U approaches, ranges of low hills, of rounded, dome-like shapes appear in the distance; at times, the ridges seem to rise into spires and finials. Such is the effect that the experience of entering the site is like stepping into a zone where the human and nonhuman echo each other with an uncanny resonance; the connection between built form and landscape seems to belong to a dimension other than the visual; it is like that of sympathetic chords in music. The echoes reach into the interiors of the monuments, which, with their openings and pathways, their intricate dappling of light and shadow, and their endless iterations of images, seem to aspire to be forests of stone.

In *How Forests Think*, the anthropologist Eduardo Kohn suggests that 'forms' – by which he means much more than shapes or visual metaphors – are one of the means that enable our surroundings to think through us.

But how, we might ask, can any question of thought arise in the absence of language? Kohn's answer is that to imagine these possibilities we need to move beyond language. But to what? Merely to ask that question is to become aware of the multiple ways in which we are constantly engaged in patterns of communication that are not linguistic: as, for example, when we try to interpret the nuances of a dog's bark; or when we listen to patterns of birdcalls; or when we try to figure out what exactly is portended by a sudden change in the sound of the wind as it blows through trees. None of this is any less demanding, or any less informative, than, say, listening to the news on the radio. We do these things all the time – we could not stop doing them if we tried – yet we don't think of them as communicative acts. Why? Is it perhaps because the shadow of language interposes itself, preventing us from doing so?

It isn't only the testimony of our ears that is blocked in this way, but also that of our eyes, for we often communicate with animals by means of gestures that require interpretation – as, for example, when I wave my hands to shoo away a crow. Nor does interpretation necessarily demand a sense of hearing or sight. In my garden, there is a vigorously growing vine that regularly attempts to attach itself to a tree, several meters away, by 'reaching' out to it with a tendril. This is not done randomly, for the tendrils are always well aimed and they appear at exactly those points where the vine does actually stand a chance of bridging the gap: if this were a human, we would say that she was taking her best shot. This suggests to me that the vine is, in a sense, 'interpreting' the stimuli around it, perhaps the shadows that pass over it or the flow of air in its surroundings

Whatever those stimuli might be, the vine's 'reading' of them is clearly accurate enough to allow it to develop an 'image' of what it is 'reaching' for; something not unlike 'heat-imaging' in weapons and robots.

To think like a forest, then, is, as Kohn says, to think in images. And the astonishing profusion of images in Mrauk-U, most of which are of the Buddha in the *bhumisparshamudra*, with the tip of the middle finger of his right hand resting on the earth, serves precisely to direct the viewer away from language toward all that cannot be 'thought' through words.

These possibilities have, of course, been explored by people in many cultures and in many eras – in fact, everywhere perhaps except within the modern academy. What, then, is to be made of the fact that such possibilities have now succeeded also in broaching the boundaries of the one sphere from which they were excluded? Could it be said, extending Kohn's argument, that this synchronicity confirms that the Anthropocene has become our interlocutor, that it is indeed thinking 'through' us? Would it follow, then, on the analogy of Kohn's suggestion in relation to forests, that to think about the Anthropocene will be to think in images, that it will require a departure from our accustomed logocentricism? Could that be the reason why television, film, and the visual arts have found it much easier to address climate change than has literary fiction? And if that is so, then what does it imply for the future of the novel?

It is possible, of course, to construct many different genealogies for the deepening logocentricism of the last several centuries. But the one point where all those lines of descent converge is the invention of print technology, which moved the logocentricism of the Abrahamic religions in general, and the Protestant Reformation in particular, onto a new plane. So much so that Ernest Gellner was able to announce in 1964, 'The humanist intellectual is, essentially, an expert on the written word.'

Merely to trace the evolution of the printed book is to observe the slow but inexorable excision of all the pictorial elements that had previously existed within texts: illuminated borders, portraits, coloring, line drawings, and so on. This pattern is epitomized by the career of the novel, which in the eighteenth and nineteenth centuries often included frontispieces, plates, and so on. But all of these elements gradually faded away, over the course of the nineteenth and early twentieth centuries, until the very word *illustration* became a pejorative, not just within fiction but in all the arts. It was as if every doorway and window that might allow us to escape the confines of language had to be slammed shut, to make sure that humans had no company in their dwindling world but their own abstractions and concepts. This, indeed, is a horizon within which every advance is achieved at the cost of 'making the world more unlivable.'

But then came a sea change: with the Internet we were suddenly back in a time when text and image could be twinned with as much facility as in an illuminated manuscript. It is surely no coincidence that images too began to seep back into the textual world of the novel; then came the rise of the graphic novel – and it soon began to be taken seriously.

So if it is the case that the last, but perhaps most intransigent way the Anthropocene resists literary fiction lies ultimately in its resistance to language itself, then it would seem to follow that new, hybrid forms will emerge and the act of reading itself will change once again, as it has many times before.

Notes

3 ignorance to knowledge: 'Recognition ... is a change from ignorance to knowledge, disclosing either a close relationship or enmity, on the part of people marked out for good or bad fortune.' Aristotle, *Poetics*, tr. Malcolm Heath (London: Penguin 1996), 18.

3 lies within oneself: In the phrasing of Giorgio Agamben, the philosopher, these are moments in which potentiality turns 'back upon itself to give itself to itself' (*Homo Sacer: Sovereign Power and Bare Life*, tr. Daniel Heller-Roazen [Stanford, CA: Stanford University Press, 1998], 46).

6 genre of science fiction: Barbara Kingsolver's *Flight Behavior* and Ian McEwan's *Solar*, both of which were widely reviewed by literary journals, are rare exceptions.

7 feedback loop: In Gavin Schmidt and Joshua Wolfe's definition: 'The concept of feedback is at the heart of the climate system and is responsible for much of its complexity. In the climate everything is connected to everything else, so when one factor changes, it leads to a long chain of changes in other components, which leads to more changes, and so on. Eventually, these changes end up affecting the factor that instigated the initial change. If this feedback amplifies the initial change, it's described as positive, and if it dampens the change, it is negative.' See *Climate Change: Picturing the Science*, ed. Gavin Schmidt and Joshua Wolfe (New York: W. W. Norton, 2008), 11.

7 wild has become the norm: Lester R. Brown writes, 'climate instability is becoming the new norm.' See *World on the Edge: How to Prevent Environmental and Economic Collapse* (New York: W. W. Norton, 2011), 47.

8 'stories our civilization tells itself': See dark-mountain.net; and see also John H. Richardson, 'When the End of Human Civilization Is Your Day Job,' *Esquire*, July 7, 2015.

8 era of the Anthropocene: Dipesh Chakrabarty, 'The Climate of History: Four Theses,' *Critical Inquiry* 35 (Winter 2009).

8 'processes of the earth': The quote is from Naomi Oreskes, 'The Scientific Consensus on Climate Change: How Do We Know We're Not Wrong,' in *Climate Change: What It Means for Us, Our Children and Our Grandchildren*, ed. Joseph F. C. DiMento and Pamela Doughman (Cambridge, MA: MIT Press, 2007). For a discussion of the genealogy of the concept of the Anthropocene, see Paul J. Crutzen, 'Geology of Mankind,' *Nature* 415 (January 2002): 23; and Will Steffen, Jacques Grinevald, Paul Crutzen, and John McNeill, 'The Anthropocene: Conceptual and Historical Perspectives,' *Philosophical Transactions of the Royal Society* 369 (2011): 842–67.

9 the wind in our hair: Stephanie LeMenager calls this 'the road-pleasure complex' in *Living Oil: Petroleum Culture in the American Century* (Oxford: Oxford University Press, 2014), 81.

11 Bangkok uninhabitable: Cf. James Hansen: "Parts of [our coastal cities] would still be sticking above the water, but you couldn't live there.' http://www.thedailybeast.com/articles/2015/07/20/climate-seer-james-hansen-issues-his-direst-forecast-yet.html.

11 Great Derangement: As the historian Fredrik Albritton Jonsson notes, if we consider the transgression of the

'planetary boundaries that are necessary to maintain the Earth system "in a Holocene-like state" . . . our current age of fossil fuel abundance resembles nothing so much as a giddy binge rather than a permanent achievement of human ingenuity' ('The Origins of Cornucopianism: A Preliminary Genealogy,' *Critical Historical Studies*, Spring 2014, 151).

15 meteorological history: The only part of the Indian subcontinent where tornadoes occur frequently is in the Bengal Delta, particularly Bangladesh. Cf. Someshwar Das, U. C. Mohanty, Ajit Tyagi, et al., 'The SAARC Storm: A Coordinated Field Experiment on Severe Thunderstorm Observations and Regional Modeling over the South Asian Region,' *American Meteorological Society*, April 2014, 606.

17 'being aware of it': Ian Hacking, *The Emergence of Probability* (Cambridge: Cambridge University Press, 1975), Kindle edition, loc. 194.

18 'into the foreground': Franco Moretti, 'Serious Century: From Vermeer to Austen,' in *The Novel, Volume 1*, ed. Franco Moretti (Princeton, NJ: Princeton University Press, 2006), 372.

18 regime of thought and practice: Cf. Giorgio Agamben on Carl Schmitt, 'the true life of the rule is the exception,' in *Homo Sacer*, tr. Daniel Heller-Roazen, 137.

19 'pictures of Bengali life': Bankim Chandra Chatterjee, 'Bengali Literature,' first published anonymously in *Calcutta Review* 104 (1871). Digital Library of India: http://en.wikisource.org/wiki/ Bengali_Literature.

19 early 1860s: See also my essay, 'The March of the Novel through History: The Testimony of My Grandfather's Bookcase,' in the collection *The Imam and the Indian* (New Delhi: Ravi Dayal, 2002).

20 'blond cornfields': Gustave Flaubert, *Madame Bovary*, tr. Eleanor Marx-Aveling (London: Wordsworth Classics, 1993), 53.

20 'no miracles at all': Franco Moretti, *The Bourgeois* (London: Verso, 2013), 381. There is an echo here of Carl Schmitt's observation: 'The idea of the modern constitutional state triumphed together with deism, a theology and metaphysics that banished the miracle from the world . . . The rationalism of the Enlightenment rejected the exception in every form' *(Political Theology: Four Chapters on the Concept of Sovereignty* [University of Chicago Press, 2005], 36–37).

21 'change in the present': Spencer R. Weart, *The Discovery of Global Warming* (Cambridge, MA: Harvard University Press, 2003), 9.

21 'does not make leaps': Stephen Jay Gould, *Time's Arrow, Time's Cycle: Myth and Metaphor in the Discovery of Geological Time* (Cambridge, MA: Harvard University Press, 1987), 173.

21 jump, if not leap: The theory of punctuated equilibrium, as articulated by Stephen Jay Gould and Niles Eldredge, proposed 'that the emergence of new species was not a constant process but moved in fits and starts: it was not gradual but punctuated.' See John L. Brooke, *Climate Change and the Course of Global History: A Rough Journey* (New York: Cambridge University Press, 2014), 29.

21 '"both and neither"': Gould, *Time's Arrow, Time's Cycle*, 191.

21 'short-lived cataclysmic events': http://geography.about.com/od/physicalgeography/a/uniformitarian.htm.

22 'immaterial and supernatural agents': Gould, *Time's Arrow, Time's Cycle*, 108–9.

22 'victim with her cold beams': Chatterjee, 'Bengali Literature.'

23 'nightingales in shady groves': Flaubert, *Madame Bovary*, 28.

24 'resent its interference': Chatterjee, 'Bengali Literature.'

24 'reigned supreme': Brooke, *Climate Change and the Course of Global History*.

24 'events in the stars': Quoted in Gould, *Time's Arrow, Time's Cycle*, 176.

24 'covers of popular magazines': Elizabeth Kolbert, *The Sixth Extinction: An Unnatural History* (New York: Henry Holt, 2014), 76. See also Jan Zalasiewicz and Mark Williams, *The Goldilocks Planet: The Four Billion Year Story of Earth's Climate* (Oxford: Oxford University Press, 2012), Kindle edition, loc. 3042, and Gwynne Dyer, *Climate Wars: The Fight for Survival as the World Overheats* (Oxford: Oneworld Books, 2010), Kindle edition, loc. 3902.

25 'basis of intelligibility': Gould, *Time's Arrow, Time's Cycle*, 10.

25 ' "carry me with you!" ': Flaubert, *Madame Bovary*, 172–73.

27 recorded meteorological history: Adam Sobel, *Storm Surge: Hurricane Sandy, Our Changing Climate, and Extreme Weather of the Past and Future* (New York: HarperCollins, 2014), Kindle edition, locs. 91–105.

27 its impacts: Ibid., locs. 120, 617–21.

28 'faraway places': Ibid., loc. 105.

28 'possible in Brazil': Mark Lynas, *Six Degrees: Our Future on a Hotter Planet* (New York: HarperCollins, 2008), 41.

29 named the 'catastophozoic': Kolbert, *The Sixth Extinction*, 107.

29 'the long emergency' and 'Penumbral Period': David Orr, *Down to the Wire: Confronting Climate Collapse* (Oxford: Oxford University Press, 2009), 27–32; and Naomi Oreskes and Erik M. Conway, *The Collapse of Western Civilization: A View from the Future* (New York: Columbia University Press, 2014), 4.

29 'extremes of heat and cold': Geoffrey Parker, *Global Crisis: War, Climate Change, and Catastrophe in the Seventeenth Century* (New Haven, CT: Yale University Press, 2013), Kindle edition, loc. 17574.

31 were killed by tigers: In his book, *The Royal Tiger of Bengal: His Life and Death* (London: J. and A. Churchill, 1875), Joseph Fayrer records that between 1860 and 1866 4,218 people were killed by tigers in Lower Bengal.

32 this fearsome sight: Amitav Ghosh, *The Hungry Tide* (New York: Houghton Mifflin Harcourt, 2005).

33 'feels it generally' Martin Heidegger, *Existence and Being*, intro. Werner Brock, tr. R. F. C. Hull and Alan Crick (Washington, DC: Gateway Editions, 1949), 336.

33 'something uncanny': Timothy Morton, *Hyperobjects* (Minneapolis: University of Minnesota Press, 2013), Kindle edition, loc. 554.

33 'menace and uncertainty': George Marshall, *Don't Even Think about It: Why Our Brains Are Wired to Ignore Climate Change* (New York: Bloomsbury, 2014), 95.

34 processes of thought: Cf. Eduardo Kohn, *How Forests Think: Toward an Anthropology beyond the Human* (Berkeley: University of California Press, 2013).

36 relationship with the nonhuman: Cf. Michael Shellenberger and Ted Nordhaus, 'The Death of Environmentalism: Global Warming Politics in a Post-Environmental World' (Oakland, CA: Breakthrough Institute, 2007): 'The concepts of "nature" and "environment" have been thoroughly deconstructed. Yet they retain their mythic and debilitating power within the environmental movement and the public at large' (12).

36 'post-natural world': Bill McKibben, *The End of Nature* (New York: Random House, 1989), 49.

43 tides and the seasons: Anuradha Mathur and Dilip da Cunha make this point at some length in their excellent book, *SOAK: Mumbai in an Estuary* (New Delhi: Rupa Publications, 2009).

43 and on Salsette: I am grateful to Rahul Srivastava, the urban theorist and cofounder of URBZ (http://urbz.net/about/people/), for this insight.

43 a chest of tea: Bennett Alan Weinberg and Bonnie K. Bealer, *The World of Caffeine: The Science and Culture of the World's Most Popular Drug* (New York: Routledge, 2000), 161.

43 'milieu of colonial power': Anuradha Mathur and Dilip da Cunha, *SOAK*, 47.

44 their colonial origins: The British geographer James Duncan describes the colonial city as a 'political tract written in space and carved in stone. The landscape was part of the practice of power.' Quoted in Karen Piper, *The Price of Thirst: Global Water Inequality and the Coming Chaos* (Minneapolis: University of Minnesota Press, 2013), Kindle edition, loc. 3168.

44 'an island once': Govind Narayan, *Govind Narayan's Mumbai: An Urban Biography from 1863*, tr. Murali Ranganathan (London: Anthem Press, 2009), 256. I am grateful to Murali Ranganathan for clarifying many issues relating to the topography of Mumbai.

44 'concentration of risk': Cf. Aromar Revi, 'Lessons from the Deluge,' *Economic and Political Weekly* 40, no. 36 (September 3–8, 2005): 3911–16, 3912.

45 cyclonic activity: A 2010 report published jointly by the India Meteorological Department and National Disaster Management Authority places the coastal districts of the India's western states in the lowest category of proneness to cyclones (table 9).

46 west coast of India: Earthquakes of 5.8 and 5.0 magnitude were recorded in the Owen fracture zone on October 2, 2013, and November 12, 2014. For details, see http://dynamic.pdc.org/snc/prod/40358/rr.html & http://www.emsc-csem.org/Earthquake/earthquake.php?id=408320.

46 'NW Indian Ocean': M. Fournier, N. Chamot-Rooke, M. Rodriguez, et al., 'Owen Fracture Zone: The Arabia–India Plate Boundary Unveiled,' *Earth and Planetary Science Letters* 302 nos. 1–2 (February 1, 2011): 247–52.

47 after the monsoons: Hiroyuki Murakami et al., 'Future Changes in Tropical Cyclone Activity in the North Indian Ocean Projected by High Resolution MRI-AGCMs,' *Climate Dynamics* 40 (2013): 1949-68, 1949.

47 region's wind patterns: Amato T. Evan, James P. Kossin, et al., 'Arabian Sea Tropical Cyclones Intensified by Emissions of Black Carbon and Aerosols,' *Nature* 479 (2011): 94–98.

49 'minor cyclonic storms': *Gazetteer of Bombay City and Island*, Vol. I (1909), 96. I am grateful to Murali Ranganathan for providing me with this reference.

49 'end of all things': Quoted ibid., 97.

49 'persons were killed': Ibid., 98.

49 people were killed: Ibid., 99.

49 'number and intensity': Ibid.

50 intensity scale: On the Saffir-Simpson hurricane intensity scale, wind speeds of 75 mph are the benchmark for a Category 1 hurricane. In the Tropical Cyclone Intensity Scale used by the India Meteorological Department, any storm with wind speeds of over 39 kmph counts as a 'cyclonic storm,' hence this storm was named Cyclone Phyan.

51 single day: R. B. Bhagat et al., 'Mumbai after 26/7 Deluge: Issues and Concerns in Urban Planning,' *Population and Environment* 27, no. 4 (March 2006): 337–49, 340.

51 estuarine location: I am deeply grateful to Rahul Srivastava, Manasvini Hariharan, Apoorva Tadepalli, and the team at URBZ for their help with the research for this section.

51 filth-clogged ditches: In 'Drainage Problems of Brihan Mumbai,' B. Arunachalam provides a concise account of how Mumbai's hydrological systems have been altered over time (*Economic and Political Weekly* 40, no. 36 (September 3–9, 2005]: 3909–11, 3909).

51 absorptive ability: Cf. Vidyadhar Date, 'Mumbai Floods: The Blame Game Begins,' *Economic and Political Weekly* 40, no. 34 (August 20–26, 2005): 3714–16, 3716; and Ranger et al., 'An Assessment of the Potential Impact of Climate Change on Flood Risk in Mumbai,' *Climate Change* 104 (2011): 139–67, 142, 146; see also R. B. Bhagat et al., 'Mumbai after 26/7 Deluge,' 342.

52 1.5 million: P. C. Sehgal and Teki Surayya, 'Innovative Strategic Management: The Case of Mumbai Suburban Railway System,' *Vikalpa* 36, no. 1 (January–March 2011): 63.

52 knocked out as well: Aromar Revi, 'Lessons from the Deluge,' 3913.

53 suffered damage: The paragraphs above are based largely on the *Fact Finding Committee on Mumbai Floods, Final Report*, vol. 1, 2006, 13–15.

53 fishing boat: Vidyadhar Date, 'Mumbai Floods,' 3714.

53 trapped by floodwaters: Aromar Revi, 'Lessons from the Deluge,' 3913.

53 homes to strangers: Cf. Carsten Butsch et al., 'Risk Governance in the Megacity Mumbai/India – A Complex Adaptive System Perspective,' *Habitat International* (2016), http://dx.doi.org/10.1016/j.habitatint.2015.12.017, 5.

53 'of the partition': Aromar Revi, 'Lessons from the Deluge,' 3912.

53 even the courts: See Ranger et al., 'An Assessment of the Potential Impact of Climate Change on Flood Risk in Mumbai,' 156.

53 swamped by floodwaters: Carsten Butsch et al. note that while many improvements have been made to Mumbai's warning systems and disaster management practices, 'there are also doubts about Mumbai's disaster preparedness. First some of the infrastructures created, are not maintained as good practice would demand; second, many of the measures announced have not been finalized (especially the renovation of the city's water system) and third, informal practices prohibit planning and applying measures.' ('Risk Governance in the Megacity Mumbai/India,' 9–10).

54 in recent years: Because of emergency measures the death toll of the 2013 Category 5 storm, Cyclone Phailin, was only a few dozen. See the October 14, 2013, CNN report, 'Cyclone Phailin: India Relieved at Low Death Toll.'

54 planning for disasters: Ranger et al. observe that while the Mumbai administration's risk-reducing measures are commendable 'they do not appear to consider the potential impacts of climate change on the long-term planning horizon.' ('An Assessment of the Potential Impact of Climate Change on Flood Risk in Mumbai,' 156).

54 'post-disaster response': Friedemann Wenzel et al., 'Megacities – Megarisks,' *Natural Hazards* 42 (2007): 481–91, 486.

54 disasters of this kind: The Municipal Corporation of Great
    Mumbai's booklet *Standard Operating Procedures for Disas-*
    ter Management Control (available at http://www.mcgm.
    gov.in/irj/portalapps/com.mcgm.aDisasterMgmt/docs/
    MCGM_SOP.pdf) is explicitly focused on floods and
    makes no mention of Cyclones. Cyclones are mentioned
    only generically in the Municipal Corporation's 2010 publi-
    cation *Disaster Risk Management Master Plan: Legal and*
    Institutional Arrangements; Disaster Risk Management in
    Greater Mumbai, and that too mainly in the context of dir-
    ectives issued by the National Disaster Management
    Authority, which was established by the country's Disaster
    Management Act of 2005. The *Maharashtra State Disaster*
    Management Plan (draft copy) is far more specific, and it
    includes a lengthy section on cyclones (section 10.4) and
    the following recommendation: 'Evacuate people from
    unsafe buildings/structures and shift them to relief
    camps/sites.' However, its primary focus is on rural areas,
    and it does not make any reference (probably for jurisdic-
    tional reasons) to a possible evacuation of Mumbai (the
    plan is available here: http://gadchiroli.nic.in/pdf-files/
    state-disaster.pdf). The *Greater Mumbai Disaster Manage-*
    ment Action Plan: Risk Assessment and Response Plan, vol. 1,
    does recognize the threat of cyclones, and even lists the
    areas that may need to be evacuated (section 2.8). But this
    list accounts for only a small part of the city's population;
    the plan does not provide for the possibility that an
    evacuation on a much larger scale, involving most of the
    city's people, may be necessary. The plan is available
    here: http://www.mcgm.gov.in/irj/portalapps/com.mcgm.
    aDisasterMgmt/docs/Volume%201%20(Final).pdf.

55 projects are located: According to an article published in the
Indian Express on April 30, 2015, '60 sea-front projects, mostly
super luxury residences,' were waiting for clearance 'along
Mumbai's western shoreline,' http://indianexpress.com/
article/cities/mumbai/govt-forms-new-panel-fresh-hope-
for-117-stalled-crz-projects/. The Maharashtra government
is also opening many unbuilt sea-facing areas, like the city's
old salt pans, to construction (see *The Hindu's Business Line* of
August 22, 2015: http://m.thehindubusinessline.com/news/
national/salt-pan-lands-in-mumbai-to-be-used-for-
development-projects/article7569641.ece).

56 corrugated iron: Carsten Butsch et al., 'Risk Governance
in the Megacity Mumbai/India,' 5.

56 Arabian Sea: Cf. C. W. B. Normand, *Storm Tracks in the
Arabian Sea*, India Meteorological Department, 1926. I am
grateful to Adam Sobel for this reference.

58 city as well: During the 2005 deluge 'The waterlogging
lasted for over seven days in parts of the suburbs and the
flood water level had risen by some feet in many built-up
areas.' B. Arunachalam, 'Drainage Problems of Brihan
Mumbai,' 3909.

58 illness and disease: See Carsten Butsch et al., 'Risk Governance in the Megacity Mumbai/India,' 4.

58 forty thousand beds: Cf. Municipal Corporation of Greater
Mumbai's *City Development Plan*, section on 'Health' (9.1;
available here: http://www.mcgm.gov.in/irj/go/km/docs/
documents/MCGM%20Department%20List/
City%20Engineer/Deputy%20 City%20Engineer%20(Plan-
ning%20and%20Design)/City%20 Development%20Plan/
Health.pdf).

58 urban limits: Aromar Revi, 'Lessons from the Deluge,'
3912.

59 rising seas: Natalie Kopytko, 'Uncertain Seas, Uncertain Future for Nuclear Power,' *Bulletin of the Atomic Scientists* 71, no. 2 (2015): 29–38.

59 'safety risks': Ibid., 30–31.

60 models predict: 'All the models are indicating an increase in mean annual rainfall as compared to the observed reference mean of 1936 mm, and the average of all the models in 2350 mm [by 2071–2099].' Arun Rana et al., 'Impact of Climate Change on Rainfall over Mumbai using Distribution-Based Scaling of Global Climate Model Projections,' *Journal of Hydrology: Regional Studies* 1 (2014): 107–28, 118. See also Dim Coumou and Stefan Rahmstorf, 'A Decade of Weather Extremes,' *Nature Climate Change* 2 (July 2012): 491-96: 'Many lines of evidence . . . strongly indicate that some types of extreme weather event, most notably heatwaves and precipitation extremes, will greatly increase in a warming climate and have already done so' (494).

60 become uninhabitable: Aromar Revi notes: 'There is a clear need to rationalize land cover and land use in Greater Mumbai in keeping with rational ecological and equitable economic considerations . . . The key concern here is that developers' interests do not overpower "public interest," that the rights of the poor are upheld; else displacement from one location will force them to relocate to another, often more risk-prone location' ('Lessons from the Deluge,' 3914).

60 threatened neighborhoods: *Climate Risks and Adaptation in Asian Coastal Megacities: A Synthesis Report*, World Bank, 2010 (available at file:///C:/Users/chres/Desktop/Current/research/coastal_megacities_fullreport.pdf). The report includes a ward-by-ward listing of the areas of Kolkata that are most vulnerable to climate change (88).

61 'below this point': http://www.nytimes.com/20u/04/21/ world/ asia/2istones.html.

64 with the 'sublime': Cf. William Cronon, 'The Trouble with Wilderness; or Getting Back to the Wrong Nature,' in *Uncommon Ground: Rethinking the Human Place in Nature*, ed. William Cronon (New York: W. W. Norton, 1995), 69– 90: 'By the second half of the nineteenth century, the terrible awe that Wordsworth and Thoreau regarded as the appropriately pious stance to adopt in the presence of their mountaintop God was giving way to a much more comfortable, almost sentimental demeanor' (6).

65 they had caused: Cf. A. K. Sen Sarma, 'Henry Piddington (1797–1858): A Bicentennial Tribute,' in *Weather* 52, no. 6 (1997): 187–93.

66 'five to fifteen feet': Henry Piddington, 'A letter to the most noble James Andrew, Marquis of Dalhousie, Governor-General of India, on the storm wave of the cyclones in the Bay of Bengal and their effects in the Sunderbunds, Baptist Mission Press' (Calcutta, 1853). Quoted in A. K. Sen Sarma, 'Henry Piddington (1797–1858): A Bicentennial Tribute.'

69 'stretches of farmland': Adwaita Mallabarman, *A River Called Titash*, tr. Kalpana Bardhan (Berkeley: University of California Press, 1993), 16–17.

69 'geography books': Ibid., 12.

70 'Flower-Fruit Mountain': *The Journey to the* West, tr. and ed. Anthony C. Yu (Chicago: University of Chicago Press, 1977.

72 'had to be evacuated': David Lipset, 'Place in the Anthro-pocene: A Mangrove Lagoon in Papua New Guinea in the Time of Rising Sea-Levels,' *Hau: Journal of Ethnographic Theory* 4, no. 3 (2014): 215–43, 233.

72 'inhuman nature': Henry David Thoreau, *In the Maine Woods* (1864).

73 likes and dislikes: Julie Cruikshank, *Do Glaciers Listen? Local Knowledge; Colonial Encounters and Social Imagination* (Vancouver: University of British Columbia Press, 2005), 8.

73 'living and non-living': Julia Adeney Thomas, 'The Japanese Critique of History's Suppression of Nature,' *Historical Consciousness, Historiography and Modern Japanese Values*, International Symposium in North America, International Research Center for Japanese Studies, Kyoto, Japan, 2002, 234.

74 'never saw an ape': Quoted by Giorgio Agamben in *The Open: Man and Animal*, tr. Kevin Attell (Stanford, CA: Stanford University Press, 2004), Kindle edition, loc. 230.

74 'words and texts': Michael S. Northcott, *A Political Theology of Climate Change* (Cambridge: Wm. B. Eerdmans Publishing, 2013), 34. See also Lynn White, 'The Historical Roots of Our Ecological Crisis,' *Science* 155 (1967): 'Christianity is the most anthropocentric religion the world has seen' (1205).

76 in recorded history: Alexander M. Stoner and Andony Melathopoulos, *Freedom in the Anthropocene: Twentieth-Century Helplessness in the Face of Climate Change* (New York: Palgrave, 2015), 10.

76 'without a Summer': Cf. Michael E. Mann, *The Hockey Stick and the Climate Wars* (New York: Columbia University Press, 2012), 39; Gillen D'Arcy Wood, '1816, the Year without a Summer,' *BRANCH: Britain, Representation and Nineteenth-Century History*, ed. Dino Franco Felluga, extension of *Romanticism and Victorianism on the Net* (http://www.branchcollective.org/); and Gillen D'Arcy

Wood, *Tambora: The Eruption That Changed the World* (Princeton, NJ: Princeton University Press, 2015).

77 John Polidori: Fiona MacCarthy, *Byron: Life and Legend* (New York: Farrar, Strauss and Giroux, 2002), 292.

77 'amid the darkness': Quoted by Gillen D'Arcy Wood, '1816, the Year without a Summer'; see also John Buxton, *Byron and Shelley: The History of a Friendship* (London: Macmillan, 1968), 10.

77 'August Darvell': Fiona MacCarthy, *Byron*, 292.

78 'vital warmth': Quoted by John Buxton, *Byron and Shelley*, 14.

78 'as we choose': Geoffrey Parker, *Global Crisis*, loc. 17871.

83 'umbrella': Margaret Atwood, *In Other Worlds: SF and the Human Imagination* (New York: Nan A. Talese/Doubleday, 2011).

85 'defend this autonomy': Timothy Mitchell, *Carbon Democracy: Political Power in the Age of Oil* (London: Verso, 2011), Kindle edition, loc. 474.

85 First World War: Ibid., locs. 430, 578.

85 transportation and distribution: Ibid., locs. 680–797: 'Whereas the movement of coal tended to follow dendritic networks, with branches at each end but a single main channel, creating potential choke points at several junctures, oil flowed along networks that had the properties of a grid, like an electricity network, where there is more than one possible path and the flow of energy can switch to avoid blockage or overcome breakdowns' (797).

86 from coal to oil: Ibid., loc. 653.

86 'energy flows': Ibid., loc. 645.

86 substance itself is not: Stephanie LeMenager's apt summation in *Living Oil*: 'Oil has been shit and sex, the essence of entertainment' (92).

86 Sebastião Salgado: There are, however, many exceptions. For a full account, see the chapter 'The Aesthetics of Petroleum' in Stephanie LeMenager's *Living Oil*.

87 'literally inconceivable': The piece is reprinted in the nonfiction collections published under the titles *The Imam and the Indian* (New Delhi: Penguin India, 2002) and *Incendiary Circumstances* (Boston: Houghton Mifflin, 2004).

89 'historical chronicle': Leo Tolstoy, 'A Few Words Apropos of the Book *War and Peace*'

89 preceding forms: Donna Tussing Orwin, 'Introduction,' in *Tolstoy on War: Narrative Art and Historical Truth in 'War and Peace,'* ed. Rick McPeak and Donna Tussing Orwin (Ithaca, NY: Cornell University Press, 2012), 3.

90 'being an egotist': *Eight Letters from Charlotte Brontë to George Henry Lewes*, November 1847-October 1850: http://www.bl.uk/collection-items/eight-letters-from-charlotte-bront-to-george-henry-lewes-november-1847-october-1850.

90 'collective metamorphosis' Rob Nixon, *Slow Violence and the Environmentalism of the Poor* (Cambridge, MA: Harvard University Press, 2011), 87–88.

91 'Great Acceleration': Cf. Will Steffen, Jacques Grinevald, et al., 'The Anthropocene: Conceptual and Historical Perspectives,' *Philosophical Transactions of the Royal Society* 369 (2011): 842–67.

91 'produce isolation': Guy Debord, *The Society of the Spectacle*, 3rd ed., tr. Donald Nicholson-Smith (New York: Zone Books, 1994), thesis 28.

91 'as progress': Bruno Latour, *We Have Never Been Modern*, tr. Catherine Porter (Cambridge, MA: Harvard University Press, 1993), Kindle edition, loc. 1412.

92 powerful presence: As Latour notes, 'the word "modern"
is always being thrown into the middle of a fight, in a
quarrel where there are winners and losers.' Ibid., loc.
269.

92 'used-up' after all: As John Barth once suggested in 'The
Literature of Exhaustion,' in *The Friday Book: Essays and
Other Nonfiction* (Baltimore: John Hopkins University
Press, 1984).

92 avant la lettre: Thus, for example, the scientist and water
expert Peter Gleick writes, in relation to the drought in
California: 'But here is what I fear, said best by John Steinbeck in *East of Eden:* "And it never failed that during the
dry years the people forgot about the rich years, and during the wet years they lost all memory of the dry years.
It was always that way."' *(Learning from Drought: Five Pri-
orities for California*, February 10, 2014; available: http://
scienceblogs.com/significantfigures/index.php/2014/02/
io/learning-from-drought-five-priorities-for-california/).

93 global temperatures: John L. Brooke, *Climate Change and
the Course of Global History*, 551.

95 move beyond language: 'We need ... to "decolonize
thought," in order to see that thinking is not necessarily
circumscribed by language, the symbolic, or the human.'
Eduardo Kohn, *How Forests* Think: Toward an *Anthropology beyond the Human* (Berkeley: University of California
Press, 2013), Kindle edition, loc. 949. See also John Zerzan,
Running on Emptiness: The Pathology of Civilization (Los
Angeles: Feral House, 2002), ii: 'Language seems often to
close an experience, not to help ourselves be open to an
experience.'

96 has literary fiction: Sergio Fava discusses some of the visual artists who have addressed climate change in his book

Environmental Apocalypse in Science and Art: Designing *Nightmares* (London: Routledge, 2013).

97 'the written word': Quoted in Arran E. Gare, *Postmodernism and the Environmental Crisis* (London: Routledge, 1995), 21.

97 'world more unlivable': The words are Franco Moretti's from *The Bourgeois* (London: Verso, 2013), 89.